Geographic Information Systems in Ecology

METHODS IN ECOLOGY

Series Editors

J. H. LAWTON FRS
Imperial College at Silwood Park
Ascot, UK

G. E. LIKENS
Institute of Ecosystem Studies
Millbrook, USA

METHODS IN ECOLOGY

Geographic Information Systems in Ecology

CAROL A. JOHNSTON
Natural Resources Research Institute
University of Minnesota
Duluth, Minnesota, USA

b
Blackwell
Science

Editorial Offices:
Osney Mead, Oxford OX2 0EL
25 John Street, London WC1N 2BL
23 Ainslie Place, Edinburgh EH3 6AJ
350 Main Street, Malden
MA 02148 5018, USA
54 University Street, Carlton
Victoria 3053, Australia
10, rue Casimir Delavigne
75006 Paris, France

Other Editorial Offices:
Blackwell Wissenschafts-Verlag GmbH
Kurfürstendamm 57
10707 Berlin, Germany

Blackwell Science KK
MG Kodenmacho Building
7–10 Kodenmacho Nihombashi
Chuo-ku, Tokyo 104, Japan

First published 1998
Reprinted 1998, 1999

Set by Semantic Graphics, Singapore
Printed and bound in Great Britain by
MPG Books Ltd, Bodmin, Cornwall

For further information on
Blackwell Science, visit our website:
www.blackwell-science.com

DISTRIBUTORS

Marston Book Services Ltd
PO Box 269
Abingdon
Oxon OX14 4YN
(*Orders:* Tel: 01235 465500
Fax: 01235 465555)

USA
Blackwell Science, Inc.
Commerce Place
350 Main Street
Malden, MA 02148 5018
(*Orders:* Tel: 800 759 6102
781 388 8250
Fax: 781 388 8255)

Canada
Login Brothers Book Company
324 Saulteaux Crescent
Winnipeg, Manitoba R3J 3T2
(*Orders:* Tel: 204 837 2987)

Australia
Blackwell Science Pty Ltd
54 University Street
Carlton, Victoria 3053
(*Orders:* Tel: 03 9347-0300
Fax: 03 9347-5001)

A catalogue record for this title
is available from the British Library

ISBN 0-632-03859-4

Library of Congress
Cataloguing-in-Publication Data
Johnston, Carol A. (Carol Arlene)
Geographic information systems
in ecology / Carol A. Johnston.
p. cm. – (Methods in ecology)
Includes bibliographical references
and index.
ISBN 0-632-03859-4
1. Ecology. 2. Geographic information
systems.
I. Title. II. Series.
QH541.J6 1998
577'.0285—dc21 97–21890
CIP

Contents

The Methods in Ecology Series

The explosion of new technologies has created the need for a set of concise and authoritative books to guide researchers through the wide range of methods and approaches that are available to ecologists. The aim of this series is to help graduate students and established scientists choose and employ a methodology suited to a particular problem. Each volume is not simply a recipe book, but takes a critical look at different approaches to the solution of a problem, whether in the laboratory or in the field, and whether involving the collection or the analysis of data.

Rather than reiterate established methods, authors have been encouraged to feature new technologies, often borrowed from other disciplines, that ecologists can apply to their work. Innovative techniques, properly used, can offer particularly exciting opportunities for the advancement of ecology.

Each book guides the reader through the range of methods available, letting ecologists know what they could, and could not, hope to learn by using particular methods or approaches. The underlying principles are discussed, as well as the assumptions made in using the methodology, and the potential pitfalls that could occur — the type of information usually passed on by word of mouth or learned by experience. The books also provide a source of reference to further detailed information in the literature. There can be no substitute for working in the laboratory of a real expert on a subject, but we envisage this Methods in Ecology Series as being the "next best thing." We hope that, by consulting these books, ecologists will learn what technologies and techniques are available, what their main advantages and disadvantages are, when and where not to use a particular method, and how to interpret the results.

Much is now expected of the science of ecology, as humankind struggles with a growing environmental crisis. Good methodology alone never solved any problem, but bad or inappropriate methodology can only make matters worse. Ecologists now have a powerful and rapidly growing set of methods and tools with which to confront fundamental problems of a theoretical and applied nature. We hope that this series will be a major contribution towards making these techniques known to a much wider audience.

John H. Lawton
Gene E. Likens

Preface

I am not a geographer. I became frustrated as a student in the only geography course I ever took, *Small Scale Cartography*, because the maps I produced with traditional Indian ink pens were pockmarked with ink blotches, unlike the lovely artwork produced by my geographer colleagues. My calligraphic ineptitude had beneficial results, however. Being unable to produce aesthetically pleasing maps, I concentrated instead on their information content. How was the mapped information collected? What biases were induced by data collection and interpolation methods? How useful was the mapped information to scientific investigation? This development of an inquiring mind provided an excellent basis for my future use of GIS in ecology.

After completing undergraduate education, my first job involved using transparent grids and a map wheel to measure the area and perimeter of mapped wetlands. It was tedious work, better suited to machines than people. I knew there must be a better way, and resolved never to use a measurement grid again. Subsequent employment as an aerial photograph interpreter and soil surveyor provided me with a landscape perspective that would also serve as a foundation for future GIS use.

I currently direct the Natural Resources GIS Laboratory at the University of Minnesota, Duluth. My colleagues have contributed to this text both directly by providing figures and examples, and indirectly by educating me about fast-breaking topics in GIS and related technologies, about which I would remain blissfully ignorant without their constant intellectual challenge.

This text focuses on aspects of GIS and related technologies that are most useful to ecologists. Of course, some GIS concepts and operations are not specific to any discipline, so the first three chapters are fairly generic. Topography is an important driver of ecological processes, hence Chapter 4 on *Topographic operations*. *Linear operations*, covered in Chapter 5, are important to ecologists who study stream networks or animal paths. The analysis of *Temporal change*, Chapter 6, is central to ecological succession. Ecologists frequently must generate their own GIS databases from field or remotely sensed data, hence Chapters 7, 8 and 9 on *Spatial interpolation, Global Positioning Systems,* and *Remote sensing*. The integration of *Modeling and GIS*, Chapter 10, is a powerful analytical and prognostic tool that is becoming increasingly widespread. Examples given in all chapters focus on ecological applications of GIS use.

This text does not dwell on specific GIS hardware or software. It does not delve into visualization or map design. Those interested in such topics should consult equipment and software documentation, or a more cartographically oriented GIS text.

Many thanks to those who contributed material and provided constructive comments about the text: John Bonde, Paul Meysembourg, Tatiana Nawrocki, John Pastor, Phil Polzer, Deborah Pomroy-Petry, Carl Richards, Jim Salés, Gerald Sjerven, Jim Westman, and Pete Wolter. Special thanks go to my parents, Phil and Dawn Johnston, and to my husband, Boris Shmagin, who nurtured this book by providing me with writing havens and loving support.

Carol A. Johnston
Duluth, Minnesota
August 25, 1997

CHAPTER 1

Background

1.1 Introduction

New scientific and technological advances periodically arise that catalyze the development of ecology. Geographic Information Systems (GIS), integrated systems of computer hardware and software for the analysis and display of spatially distributed data, constitute this kind of keystone technological advance. GIS is a technology with a broad base of applicability, and as such, its relevance to a particular discipline may not be readily apparent. Similarly, technological capabilities that may be of paramount importance in one discipline may be trivial in another. Therefore, this text is written for ecologists. It seeks to demonstrate how GIS can be used as a **tool** in ecological research, not as an end in itself. Its focus is the analytical, rather than graphical, capabilities that can be brought to bear on ecological problems.

1.1.1 What is a GIS?

A GIS consists of computer hardware and software for entering, storing, retrieving, transforming, measuring, combining, subsetting, and displaying spatial data that have been digitized and registered to a common coordinate system. In order to perform these functions, the data entered into a GIS must include information about the spatially explicit **location** of an entity, as well as its **attributes**.

At a minimum, a GIS should be capable of the following:

- data input, editing, and management;
- data storage and retrieval;
- performing queries based on entity attributes, location, or combinations thereof;
- generating new databases based on those queries;
- producing tabular, graphical, and digital output.

Some of these capabilities (e.g. data input, output, editing, storage, and retrieval) are shared by other types of computer programs, but the ability to provide answers to geographic queries distinguishes GIS (Goodchild 1985).

A GIS differs from a map in several ways. A map is an analog depiction of the Earth's surface, whereas a GIS records spatially distributed features in numerical form. A map simultaneously depicts a variety of landscape features (e.g. topography, vegetation, road networks), whereas a GIS stores these features separately. A map is static and difficult to update, whereas a GIS **data layer** can be revised easily. A map is its own end product, whereas the end product of a GIS analysis may be a map or data. Although maps can be a form of GIS input or output, a GIS greatly enhances the versatility of mapped data due to its wealth of techniques for data manipulation and quantitative analysis.

There are two basic types of GIS, which differ in the way they store data. A **raster-based** GIS, also known as a grid- or pixel-based system, portrays features as a matrix of grid cells, each with an individual data value. A **vector-based** GIS portrays features as points, lines, and polygons. Each data structure has advantages and disadvantages, depending on the type of GIS application (see Chapter 2).

1.1.2 Why should ecologists be interested in GIS?

Geographic Information Systems are often considered the domain of geographers, a myth that tends to be perpetuated by geographers and ecologists alike. Putting aside disciplinary preconceptions, one must examine how the scientific needs are addressed by the capabilities of the technology. A GIS is best suited to analyze questions of a spatial nature, questions in which the location of a biological entity relative to other organisms or the environment influences its functioning.

Examples of ecological questions a GIS can address are:

- Where does plant community A exist?
- Where is plant community A in relation to plant community B?
- How is the distribution of plant community A related to environmental factors X, Y, and Z?
- How has the distribution of plant community A changed over time in the past?
- How will the distribution of plant community A change in the future if current environmental conditions remain the same?
- How will the distribution of plant community A change if environmental factor X is altered?
- How has the amount, location, and type of patch edge affected the movements of animal J?
- How have the territorial bounds of animal group C encroached on those of animal group D?
- How do ecosystem components X, Y, and Z contribute to material export from watershed Q?

A GIS is conventionally used to portray features on the Earth's surface, but its data management and analysis capabilities can be used with any spatially distributed data. For example, GIS has been used to analyze plant root systems photographed from behind a vertical glass plate, insect damage to leaves, and underground water movement. GIS can analyze data at a variety of spatial scales, from microscopic to global.

1.2 Related systems

GIS is related to, and often linked with, other types of software that are distinguished by data used and analyses performed. Although these systems have been designed for other purposes, they share with GIS many of the same functions, and distinctions among them have become blurred over time (Fig. 1.1).

A database management system (**DBMS**) is a computer program to organize and query data. A GIS contains DBMS functions, but a DBMS lacks the spatial query and display capabilities of a GIS. Some GIS have internal database management systems, whereas others are linked to stand-alone DBMSs (e.g. dBase, INFO, Informix, Ingress, Oracle) that allow the user to manipulate the data outside of the GIS.

Statistical analysis is an important ecological research tool, but most GIS can perform only rudimentary statistical operations (sums, means, minimum and maximum values). Therefore, ecologists commonly transfer data from a GIS into a **statistical package** for more advanced analysis, or use a software program that automatically links a GIS with a statistical package. Some statistical packages are capable of generating maps and data surfaces very similar to GIS output (SAS Institute 1990; Statistical Sciences 1991).

Computer Aided Design (**CAD**) systems were designed initially for drafting plans of technical objects, but have been used for geographic applications. Like a vector GIS, a CAD relates points, lines, and polygons

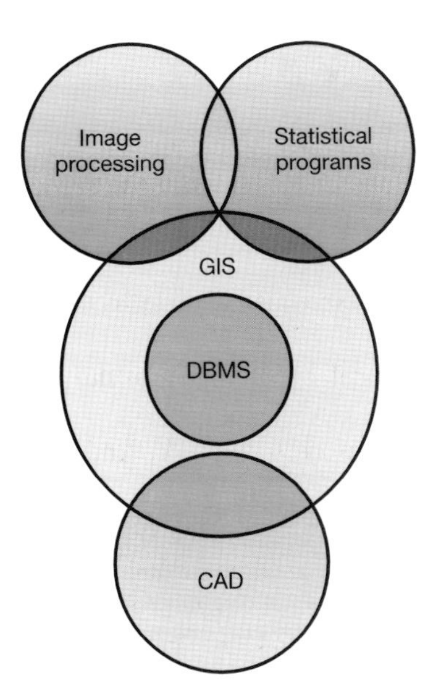

Fig. 1.1 Relationships between GIS and related computer systems.

to a spatial frame of reference. Different types of features are stored in different data layers, which then are combined to generate a final product. Unlike a GIS, however, it is very difficult to attach database attributes to specific geographic entities and automatically assign symbology on the basis of user-defined criteria (Cowen 1988).

An **image processing system** consists of software and hardware for analyzing digital images. Satellite imagery is acquired as a matrix of pixels (picture elements), each pixel containing digital spectral data for an individual cell of a gridded image. For example, each Landsat Multi-Spectral Scanner (MSS) pixel covers a ground surface area of 79 × 79 m (see Chapter 9). Image analysis systems were designed to analyze combinations of spectral data to produce a map of land surface features. Because they operate on pixels in a gridded matrix, image processing systems have many of the same capabilities as a raster GIS, but have additional programs for analyzing and classifying spectral data. Image processing systems have also been developed for analyzing medical imagery, such as MRI (magnetic resonance imaging) and CAT (computerized axial tomography) scans.

1.3 GIS development

1.3.1 Early GIS

Numerous conceptual, analytical, and technological advances in the 1960s to mid-1980s contributed to the development of GIS for ecological applications. The use of hand-drawn overlays to stack layers of thematic data may be traced back at least 100 years in the field of landscape architecture (Steinitz *et al.* 1976), but it was Ian McHarg's 1969 book on land suitability analysis, *Design with Nature*, that galvanized interest in the technique. Land suitability analysis used overlays of spatially referenced data for resource planning and management decision-making, a technique that is a central function of most GIS.

Harvard's laboratory for computer graphics was a major incubator for GIS in the early 1960s. The SYMAP program, written by Howard T. Fisher in 1963, was the prototype computer mapping system designed for spatial data analysis. SYMAP can process point and area data, interpolate a surface, and create a digital matrix (Sheehan 1979). Data surfaces can be depicted as isoline or gray scale maps, using the overprinting of line printer characters to produce suitable gray scales. SYMAP was followed by two raster systems, GRID and IMGRID, and a vector GIS, ODESSEY (Sinton & Steinitz 1969; Sinton 1977; Morehouse & Broekhuysen 1981).

Several major land use inventories conducted in the 1960s to the early 1970s served as the impetus for further GIS development. The first national scale GIS was the Canada Geographic Information System

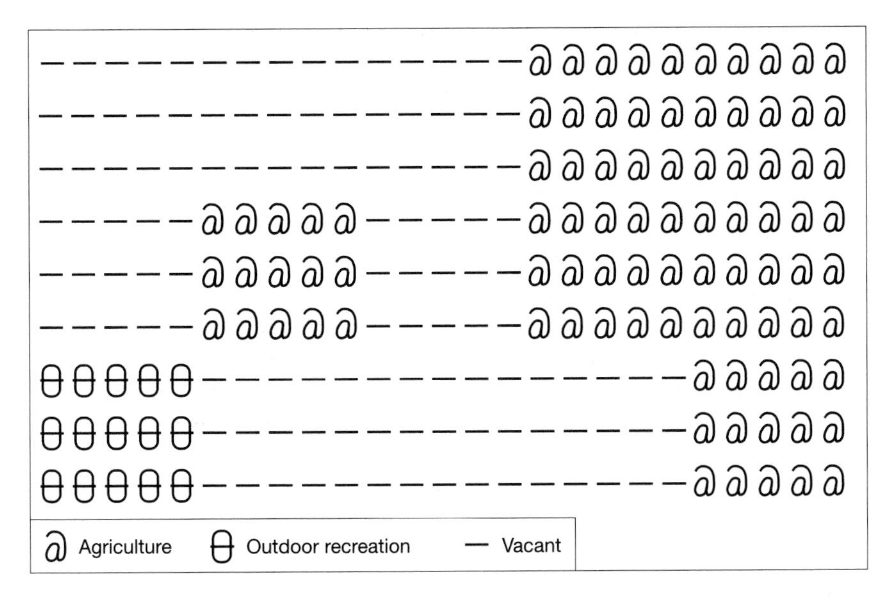

Fig. 1.2 PLANMAP output from the New York State LUNR Inventory covering a 15 km^2 area.

(CGIS), implemented in 1964 to analyze Canadian Land Inventory data (Tomlinson *et al.* 1976). This system was followed in the late 1960s by the New York State Land Use and Natural Resources (LUNR) inventory, conducted at Cornell University (Hardy & Shelton 1970) and the Minnesota Land Management Information System (Craig 1985). The LUNR Inventory stored summaries of point, line, and area data in 140 000 grid cells, each representing 1 km^2 of the state. Computer-generated LUNR inventory products included DATALIST, lists of area, linear, and point information for each cell, either as raw data or mathematically manipulated, and PLANMAP, computer-generated maps indicating the location and results of quantitative data queries (Fig. 1.2).

Some early GIS have withstood the test of time, evolving into versions that are still in use today. The CGIS has been extensively modified, and now operates as one component of the Canada Land Data Systems, an integrated group of computer-based geographic information systems (Aronoff 1989). A GIS developed in 1972 for the Minnesota Land Management Information System, EPPL1, has evolved into a PC-based raster GIS called EPPL7 (Land Management Information Center 1992). An early vector GIS also developed by the state of Minnesota, PIOS, served as the foundation for today's ARC/INFO® software*. Variants and up-

*ARC/INFO® is a registered trademark of Environmental Systems Research Institute, Inc.

dates of MAP (Tomlin 1986) and IDRISI (Eastman 1992a), two raster GIS software packages developed at U.S. universities during the mid-1980s, are still widely used for GIS training and applications. Additional information on the history of GIS is presented by Carter (1984) and Burrough (1986).

1.3.2 Database development

The onset of the computer age revolutionized the storage and management of large databases that were previously inaccessible except by tedious manual searching. Centralized computer repositories were developed by government agencies to house and manage large data sets. Locational information often was stored as part of data records, but generally as textual descriptors (e.g. county name, road intersection) rather than geographic coordinates. In the U.S., the Domestic Information Display System (DIDS) was an attempt by the Census Bureau and the National Aeronautic and Space Administration (NASA) to make such large databases more accessible. DIDS applied choroplethic mapping to data contributed by many federal agencies, but was a failure as an information system due to its inability to answer relevant spatial questions (Cowen 1983).

One of the most enduring nationwide computer databases relating to resource management and quality is the STORET database, developed by the U.S. Public Health Service for storing water quality data (Green 1964), and still maintained by the U.S. Environmental Protection Agency. Although the locational data (and sometimes the water quality data) in STORET must be carefully checked for errors, STORET remains a viable source of data for GIS analyses of water quality distribution and trends (Johnston *et al.* 1988a).

The space race that ensued after the 1957 launch of the Sputnik satellite by the U.S.S.R. also benefitted GIS-related database development. In 1972, NASA launched the Earth Resources Technology Satellite (ERTS), the first unmanned satellite specifically designed to acquire data about the Earth's resources. Although the ERTS program was experimental, its potential benefits were realized quickly by the world's nations, and about 300 individual ERTS-1 experiments were conducted in 43 U.S. states and 36 nations (Lillesand & Kiefer 1994). The ERTS program was renamed Landsat in 1975. Landsat imagery continues to provide timely, useful data for GIS applications (see Chapter 9).

1.3.3 Technological advances

Major advances in GIS development often were spurred by advances in computer hardware. Early GIS were written for mainframe computers, with data entry on punched cards and data storage on magnetic computer-compatible tapes (CCT). Disk storage was the first major technological advance to benefit GIS development. The New York State LUNR inventory utilized a computer disk storage system that provided the advantages of: "(1) Random access to any item of information in any cell, and (2) extended storage capacity far beyond foreseeable needs of the inventory" (Hardy & Shelton 1970). Although standard in contemporary computers, disk storage was a major advance over the punch card and tape storage media that were prevalent at the time.

Computers decreased in physical size while increasing in power, and GIS were written for smaller computer platforms: minicomputers and microcomputers (PCs). In his book, *Computer Mapping, Progress in the '80s*, James R. Carter (1984) writes "Now we are witnessing the invasion of the microcomputer." Carter was correct in predicting the explosion of personal computer (PC) use in the GIS arena, a trend that popularized GIS because of its affordability. The PC is the computer platform of choice for the majority of GIS users, and hundreds of GIS software packages are written for the PC (GIS World 1996).

Computer workstations, with their high processing power and large storage capacity, are becoming increasingly popular as GIS platforms. Workstations linked by network connections can share disk storage, allowing multiple users access to GIS programs and data. The windowing environment that is standard on workstations allows users to perform several operations simultaneously, and facilitates data transfer from one computer program to another. Windowing software is also now standard on PCs.

The **CD-ROM** (Compact Disk-Read Only Memory) has revolutionized the distribution of GIS databases because of its durability and large storage capacity. The storage capacity of some CD-ROMs exceeds 650 megabytes (Mb: see Section 1.4.1). Additional advantages of CD-ROM include lower reproduction costs and longer stability than magnetic media, consistent with the long-term validity of much geographic data. Most GIS software programs are now distributed on CD-ROM, and many U.S. government agencies have begun distributing large GIS data sets on this medium at minimal cost to the user.

Global Positioning Systems (**GPS**) have made it possible to collect accurate data about the *location* of organisms and field samples, a capability that was previously very difficult and expensive using conventional surveying techniques (see Chapter 8). As GPS equipment decreases in cost and complexity, it is being used increasingly by field ecologists. These locational data can then be imported and analyzed in a GIS.

1.3.4 Conceptual advances

In the mid-twentieth century, the development of mathematics appropriate for spatial problems promoted quantitative thinking about ecological pattern. Information theory was used to develop diversity indices for quantifying heterogeneity of species assemblages (Margalef 1958; Shannon & Weaver 1963; Pielou 1966) and landscapes (Godron 1966). Distance measures were devised to quantify plant community structure (Cottam & Curtis 1956; Pielou 1959, 1960). The concept of fractal dimension was introduced (Mandelbrot & Wallis 1969) and seminal works on geostatistics were published (Matheron 1965; Krige 1966). These mathematical advances ultimately contributed to the spatial analysis capabilities of contemporary GIS.

Landscape ecology emerged as a separate field in the early 1980s (Naveh 1982; Forman & Godron 1986). Landscape ecologists soon embraced GIS as an essential tool, and the International Association for Landscape Ecology formed a GIS working group in 1988. With its focus on the hetergeneous distribution of ecological resources, populations, and processes over space and time, landscape ecology has greatly benefitted from and contributed to the development of GIS.

1.3.5 Adoption and development of GIS in ecology

GIS began to be adopted by U.S. ecologists in the late 1980s. GIS was featured in a special symposium held at the 1987 Ecological Society of America Annual Meeting in Davis, California, and papers presented were published in a special issue of *Landscape Ecology* (Johnston 1990). The U.S. National Science Foundation (NSF) promoted the use of GIS in ecology by funding GIS training and facilities through its Long Term Ecological Research and Biological Facilities Center programs (Johnston 1989; Michener *et al.* 1990). A compendium edited by Cousins and co-workers (1993) illustrated advances in the use of GIS for landscape ecology in Europe and the U.S., as did a later compendium edited by Michener and co-workers (1994).

A series of international conferences on integrating GIS and environmental modeling, organized under the aegis of the U.S. National Center for Geographic Information and Analysis, has been pivotal in the exchange of knowledge and ideas about the use of GIS in ecology (Goodchild *et al.* 1993; Goodchild *et al.* 1996; NCGIA 1996). The proceedings for the 1996 conference were published as a CD-ROM and a World Wide Web home page, facilitating rapid dissemination and the use of color graphics.

GIS has become a standard tool in landscape ecology, and analytical techniques developed in this subdiscipline have become incorporated into GIS software (see Chapter 3). GIS is less widely used by field ecologists, but may become increasingly adopted with the evolution of GPS, geostatistics, and other techniques for the collection and analysis of spatially explicit field data.

1.4 Basic components of a GIS

GIS hardware consists of the computer's central processing unit (CPU) used to run the software, and various peripherals for data entry, data storage, visual display, and hard copy output. Although GIS has been developed for virtually every CPU platform, GIS operated on microcomputers (PCs) and workstations have become increasingly popular. A comprehensive review of GIS software and hardware is beyond the limits of this text, and would quickly become outdated due to constantly changing technology. GIS hardware is discussed in more detail by Antenucci *et al.* (1991), and up-to-date information on GIS products and services is published annually in the International GIS Sourcebook (e.g. GIS World 1996).

1.4.1 Bits and bytes

In digital computing, the smallest element of information is a **bit**, or binary digit, which has a value of 0 or 1. A **byte** is a group of 8 bits, for which there are 2^8 or 256 unique values. The size of GIS files is measured in megabytes (**Mb** = 10^6 bytes) or even gigabytes (**Gb** = 10^9 bytes). Data density on magnetic tapes is measured in bytes per inch (**BPI**).

Unlike the decimal number system, computers use binary notation to store numbers, in which a series of 8 bits are numbered 0 through 7, and signify as follows:

Bit:	7	6	5	4	3	2	1	0
	128s	64s	32s	16s	8s	4s	2s	units

The binary number 01110101, for example, has no 128s, one 64, one 32, one 16, no 8, one 4, no 2, and one unit, and is thus equal to the decimal number 117 ($= 64 + 32 + 16 + 4 + 1$). Early PCs had 8-bit processors, which restricted the numbers that could be stored in a GIS database to integers between 0 and 255. The development of more powerful PCs and workstations has allowed the use of 16-bit ($2^{16} = 65\,536$), 32-bit ($2^{32} = 4.295 \times 10^9$), and floating-point notation, expanding the range and precision of numbers that can be used in a GIS.

1.4.2 Data entry

GIS hardware and software are worthless to ecologists without appropriate data. Although GIS-compatible data sets are being collected by a number of federal and international agencies, they usually are more relevant to the social sciences than to ecology (see Chapter 2). Therefore, ecologists often must generate their own databases and enter them into a GIS, a time-consuming task that may be a barrier to GIS use.

Data can be entered into a GIS in a number of ways, three of which (field data entry, digitizing table, and automated scanning) are described below. Data entry by importing existing digital databases and by heads-up digitizing of imagery are discussed in Chapters 2 and 9, respectively.

Field data entry. The most basic means of entering field data is to manually key a feature's location and attributes into a computer database. Keyboard entry is relatively simple for point data, such as the location of sampling sites, but is prohibitively slow for detailed maps.

Spatially distributed field data collected with a data logger can be directly downloaded into a computer file, and related to a corresponding GIS database of sample locations. Similarly, digital location data collected with a global positioning system (GPS) can be downloaded directly into a GIS-compatible computer file (see Chapter 8). As GPS becomes more widely used in ecology, both field data and their locations will be collected simultaneously.

Digitizing table. Mapped data usually are entered into a GIS with a digitizing table. The map is placed on the table, and an operator traces the linework with a puck which electronically converts point locations and line segments into *x,y* coordinates. A digitizing table is a vector device, creating files of points and lines. When the digitized data are to be used in a raster GIS, software then must be used to assign the points and lines to grid cells. When the data are to be used in a vector GIS, the digitized file must be processed to define **topology**, the spatial relationships among a set of features. Digitizing maps is a tedious, labor-intensive process, and digitizing complex databases covering large areas can be very time-consuming.

Automated scanning. For large databases, automated scanning devices often provide a faster means of data entry than a digitizing table. Scanners can be separated into two types, those that scan lines by following them directly, and those that operate in a raster mode, breaking the map image up into tiny squares. Line-following scanners trace lines by laser beam. An operator positions the laser beam over the starting point, and attaches a label to the digitized line for later attribute assignment. Raster scanning devices include flat-bed scanners, drum scanners, and scanning cameras. A raster scanner converts a black and white or color image of a map into a finely gridded digital data file. Image processing (see Chapter 9) or line vectorization software then must be used to convert the matrix of scanned pixels into a classified, spatially referenced database. Line vectorization software converts the raster image into points and lines suitable for use in a vector GIS. Some vectorization can be done in an automated fashion, but substantial editing may be required where lines intersect or are close together (e.g. closely spaced contours on a topographic map). Line attributes and labels must be added interactively to a vectorized data layer.

Scanned input to a GIS need not always be a map. Video and digital cameras can create images of aerial photos, which then can be used as the backdrop for vector GIS data, or processed with an image analysis program to create a GIS data layer without an intermediate map. Images of objects (e.g. leaves, roots, pieces of gravel) can be created and analyzed with the same methodology used for aerial photos.

Although automated scanning has the appeal of reducing the drudgery of manual digitizing, scanning technology is still very labor-intensive. Scanned maps require careful checking and editing for erroneous line gaps and overlaps. Most maps are not drafted with digitization in mind, and although the various lines and symbols that appear on a map may be readily differentiated by the human brain, they may be indistinguishable to a computer. Continued development is needed to advance automated scanning and reduce its cost and labor intensity.

1.4.3 Data storage and dissemination

Hard disks. During processing and analysis, GIS databases are stored on a computer's magnetic hard disk. Computers with large hard disk capacity, of the order of gigabytes, are preferable for GIS use because of the large size of most GIS database files. Computers that are linked by local area networks (**LAN**) can share access to databases stored on a hard disk anywhere within the network. Computers linked remotely by the **Internet** can also share databases, with appropriate user permission, limited only by the speed of the connection.

Other magnetic media. Computer-compatible tape (e.g. 9-track tape) was the first magnetic medium used to store and transfer large GIS databases, but is now rarely used because of its bulk, slow access speed, limited storage capacity, and specialized hardware requirements. Floppy disks are a more widespread magnetic medium because floppy disk drives are standard equipment in PCs and many workstation computers, but the small storage capacity per disk (1.44 Mb) limits their use to small files. Magnetic tape cartridges are a compact storage medium, used often for archiving files and as a means of providing **back-up** storage in the event of hard disk corruption. Some types of tape cartridges are capable of storing several gigabytes of data. The drives needed to read or write to tape cartridges may be internal or external to a computer, and utilize a variety of tape sizes and recording formats. This lack of standardization complicates their use for transferring files between different GIS installations. Some cassette tape drives utilize inexpensive 8-mm tape cartridges manufactured for video cameras, which reduces the cost of data storage.

CD-ROMs. CD-ROM disks are beginning to replace magnetic media as the standard for database archival and dissemination due to their durability and large storage capacity. GIS software programs are also distributed increasingly by CD-ROM. Equipment for writing CDs is now sufficiently inexpensive (several hundred U.S. dollars) to put this medium within reach of the average GIS user. New high density CDs, called DVDs (Digital Video Disc or Digital Versatile Disc), will have from 5 to 7 Gb per side and may store as much as 20 Gb total data on one disc.

1.4.4 Presentation display

Given the power of an image to convey complex information (i.e. a picture is worth 10 000 words), visual display is an important GIS product. However, the ability of GIS to generate attractive graphics has caused some ecologists to adopt an "Emperor's New Clothes" stance on GIS, perceiving it as just another elaborate (and expensive) way to draw pretty pictures. A GIS's graphic display capability is certainly one of its most aesthetic properties, but should not be misconstrued as just a facade.

Monitors. A monitor allows the GIS user to interact with a computer, displaying commands entered and results derived. In addition to this utilitarian purpose, monitors and computer projection systems can be an important visualization device, displaying vivid, multi-color renditions of GIS databases. The advantage of using a computer projection system over still media such as view graphs or slides is that GIS operations can be displayed in **real time** (i.e. instantaneously as they occur).

A monitor is a raster device, portraying features as a fine matrix of cells. Most monitors are controlled by video cards, subject to the same binary number thresholds as any other digital device. For example, a 4-bit monitor displays only 16 different colors (2^4), an 8-bit monitor can display 256 (2^8), and a 16-bit monitor can display 65 536 (2^{16}). **RGB** monitors, which have separate inputs for red, green, and blue light, display an unlimited number of colors, and are often used in remote sensing applications (see Chapter 9).

Videotape. Special graphics cards and software can be used to link a computer with a video cassette recorder (VCR), so that a sequence of GIS-generated images and explanatory text can be recorded to videotape. The advantages of this medium are that: (i) image files which would be too large for floppy disk storage can easily be recorded to videotape; (ii) VCR technology is in widespread use, even by computer-phobics; and (iii) motion can be displayed, such as a multitemporal sequence of spatially distributed model predictions. The main disadvantages of using videotape are that it has relatively poor resolution and cannot be modified in real time.

1.4.5 Hard copy output

Ecological GIS applications place much less emphasis on map production than do cartographical GIS applications, but usually some type of hard copy output is required, regardless of the application. As with digitizing, both raster and vector hard copy output devices exist: vector devices (i.e. pen plotters) apply ink as lines, whereas raster devices apply ink as dots. However, many raster devices can accept instructions in either raster or vector format. If the hard copy output device is connected directly to the CPU that runs the GIS, a printer driver is used to generate the output. Alternatively, the GIS can be used to generate a file in a particular graphics language, such as PostScript or HPGL, that instructs a remote output device. The choice of a hard copy output device depends on one's budget, the desired media (e.g. photographic slides versus paper maps), the size and complexity of the desired output, the devices supported by the GIS used, and whether or not the output must be in color (Table 1.1).

Pen plotters. Pen plotters operate in a vector mode, drawing maps as a series of lines on paper held on a flatbed or rotating drum. Multicolor maps can be drawn using different pen colors, and linear shading patterns can be used to fill in polygons (Fig. 1.3). Plotters generate output ranging from page size to large format (61–112 cm wide) on a variety of media (e.g. paper, mylar, scribing film). The main disadvantages of plotters are that they are slow and cannot convert data to raster format, making it difficult to generate blocks of solid color. Pen plotters are also limited in the number of colors they can depict. Most have only four to eight pens, although closely spaced lines in different color combinations can give the appearance of other colors.

Impact printers. Impact printers generate hard copy with a matrix of tiny pins that strike a printer ribbon to leave ink dots on the medium behind. The earliest computer mapping systems used this rudimentary

Table 1.1 Comparison of different hard copy devices for GIS output. Plot sizes given in categories recognized by the American National Standards Institute.

Device	Relative cost	Media used	Media size	Resolution (dpi)	Number of colors
Pen plotter	Variable	Roll or cut sheet paper, mylar, scribing film	A–E	Not applicable – vector output only	4–8 pens
Impact printer	Low	Continuous feed paper	A, B	70–150	1–8 ribbons
Laser printer	Moderate	Plain paper	A	300–600	Black, color
Thermal transfer	Moderate	Treated paper	A	200–300	Continuous tone
Color copier	High	Plain paper	A, B	⩾400	Continuous tone
Ink jet plotter	Moderate	Special paper	A–E	120–720	Continuous tone
Electrostatic plotter	High	Special paper	A–E	⩾400	Continuous tone
Film recorder	Variable	Film (slides, negatives)	Usually 35 mm	~3000	Continuous tone

A-size = 8.5 × 11″, B-size = 11 × 17″, C-size = 18 × 24″, D-size = 22 × 36″, E-size = 34 × 44″.

Fig. 1.3 Hard copy map generated with a pen plotter. Courtesy of John Pastor, Natural Resources Research Institute, University of Minnesota, Duluth.

means of hard copy output (Fig. 1.2). Contemporary dot matrix printers are more advanced than the early line printers, and can use multiple color ribbons to generate up to eight colors. However, they are slow, have low resolution, and tend to produce colors with inconsistent hue, brightness, and overall clarity (Antenucci *et al.* 1991).

Laser printers. Laser printers are used commonly for word processing output, and can generate high-quality GIS output as well (Fig. 1.4). Laser printers produce images by using a laser to apply charges to a rotating drum that attracts a dry toner. Plain paper then is pressed against the drum, fusing the toner to the paper. The main limitation of laser printers for GIS applications is the small size of their output, generally 8.5 × 11″ (21 × 30 cm). Laser printers can accept digital input from both raster- and vector-based GIS, usually via a PostScript language file. Although black

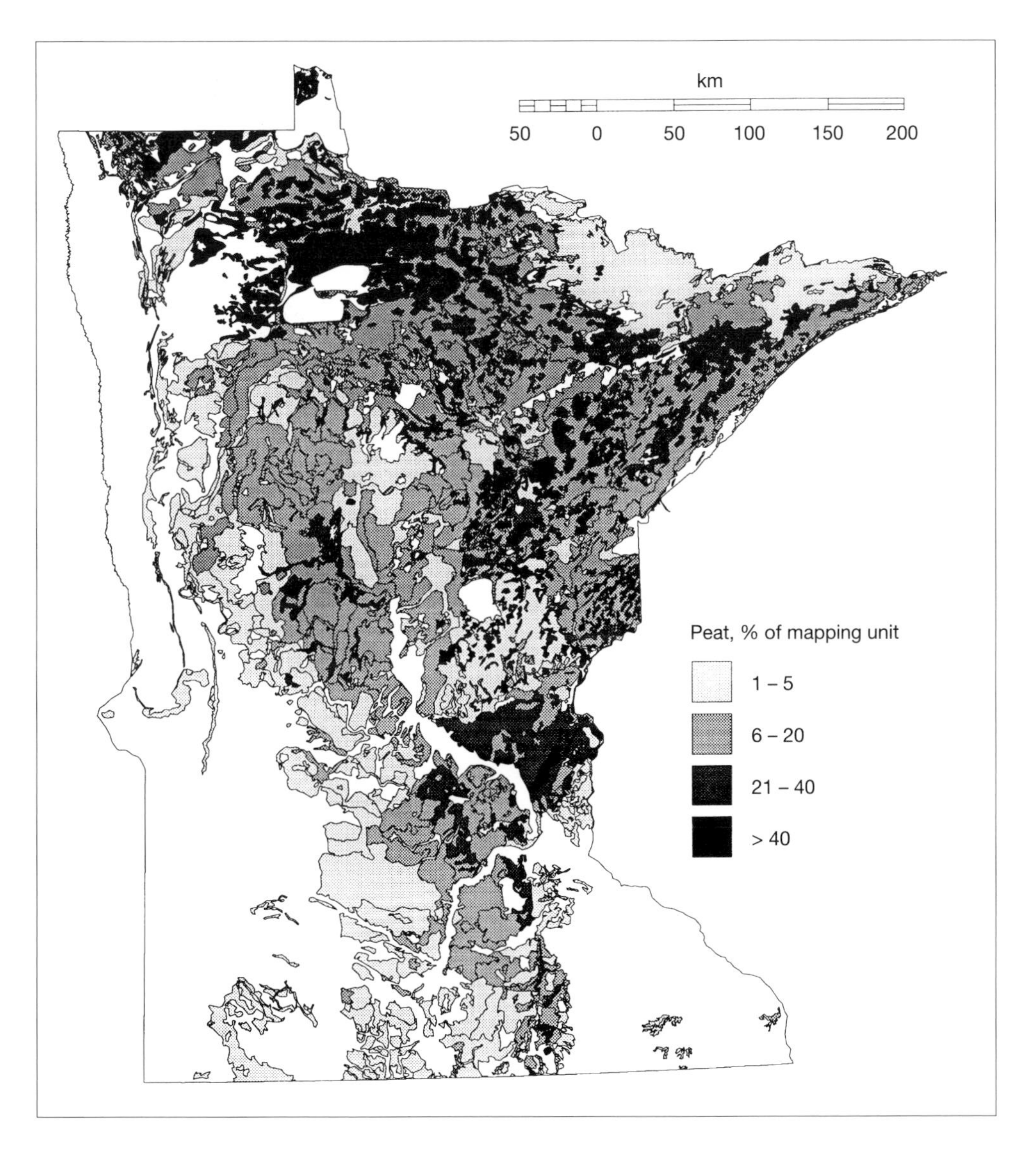

Fig. 1.4 Hard copy map generated with a black and white laser printer. Courtesy of Paul Meysembourg, Natural Resources Institute, University of Minnesota, Duluth.

and white laser printers are the norm, color laser printers are also available, albeit expensive. Speed, high resolution (⩾ 300 dots per inch), and relatively low cost make this the printer of choice for page-sized black and white plots.

Thermal transfer. Thermal transfer plotters produce high-quality color hard copy using a heating element to melt dots of waxy ink from a film substrate onto a specially treated paper. The transfer is done in three or four passes, one pass for each primary color plus black. Resolutions typically range from 200 to 300 dots per inch. They generally produce page-sized output.

Color copiers. Although more commonly used for making photocopies, some advanced color copiers can accept direct digital input, such as a PostScript file. Plain paper or other plotting medium is electrically charged by a series of small nibs, then passed through a liquid toner solution that adheres where the electrical charges were applied. Color copiers usually produce output no bigger than 11 × 17″ (28 × 43 cm), which is their primary disadvantage. Color copiers are expensive to purchase, but are becoming increasingly available in private and public sectors due to the popularity of this technology.

Ink jet plotters. Ink jet plotters force inks through small jets onto the print media, with a resolution of 120–720 dots per inch. Color ink-jet printers can produce thousands of vivid colors with varying combinations of cyan, yellow, magenta, and black inks. Like laser printers, their high resolution allows them to accept hard copy output from both raster- and vector-based GIS with the use of a suitable printer driver or a file written in a compatible graphics language (e.g. PostScript). Some ink jet plotters accept rolls of paper up to 88 cm wide, and can produce plots up to 15 m long!

Electrostatic plotters. Electrostatic plotters, in which ink is transferred to electrostatically charged areas of paper, can produce very high resolution (up to 400 dots per inch), multicolor, large format output. Depending on the complexity of the plot, electrostatic plotters can generate a product from five to more than 30 times faster than pen plotters (Antenucci *et al.* 1991). They are expensive, but produce such high-quality output that they are the plotter of choice for large format color hard copy.

Film recorders. A film recorder is a computer-driven camera system that creates a color image directly on film. Most commonly used for generating 35-mm slides, they can also be used to produce color negatives for photographic reproduction. High-resolution film recorders are available that produce color separation plates for offset printing.

1.5 Venturing into ecological GIS

Ecology is a broad field, requiring knowledge about biology, chemistry, statistics, and the environment. The use of GIS requires additional knowledge of computers, peripheral equipment, computer operating systems, data management, and GIS software. Given the difficulty of keeping current with literature and advances in ecology itself, learning to use GIS can be daunting. Ecologists should not naïvely assume that using a GIS is comparable to using a word processor or spreadsheet program. Although some GIS packages are relatively simple to install and operate,

Table 1.2 Average time required to complete a photographic-based Wisconsin Wetlands Inventory map. From Johnston *et al.* 1988b.

Task	Person-hours/36 mile^2 township	
	Excluding leave time	Including leave time
Stereoscopic photo interpretation	15.3	17.6
In-house check of photo interpretation	5.0	5.8
Field check	6.3	7.2
Drafting	27.9	32.1
In-house check of drafting	3.8	4.4
Total	58.3	67.1

high-end systems require knowledgeable and experienced operators. Qualified GIS technicians are difficult to find in an academic setting because of the high demand and pay for such individuals in the private sector. A common solution is to form an alliance with a geography department that can supply GIS equipment and expertise, but this arrangement is not always satisfactory due to the different objectives of ecological GIS.

Ecologists should be especially careful before initiating major GIS database development. Map digitizing is tedious and time-consuming, but given the importance of the source data to the rigor of the final analysis, database development should not be taken lightly. References that provide time requirements for mapping and digitizing tasks may help with project planning (Table 1.2).

The following advice should be heeded before embarking on any GIS project:

- **Keep it simple** — start with simple data and software. Although new users may be attracted to powerful, top-of-the-line GIS packages, a simpler package may be perfectly adequate.
- **Read documentation** — good manuals and on-line help are invaluable.
- **Use existing data** — where possible, use existing GIS databases rather than develop new ones.
- **Plan ahead** — a GIS analysis usually requires multiple steps which should be formulated in advance. A data management or flowcharting tool (e.g. GEOLINEUS: Lanter 1992) may be useful.
- **Keep good records** — a description of the source data and analysis performed should be written for each step in the GIS process.
- **Check results** — determine that the results obtained from a GIS procedure are logical before continuing on to the next step in the analysis.
- **Consult with experts** — before embarking on any GIS project, consult with experienced individuals for advice on database management and GIS procedures.

CHAPTER 2

GIS data

2.1 The cartographic model of data depiction

A map is a pictoral representation of Earth's features, showing their location relative to a coordinate system or other frame of reference. Humans have made maps ever since the first cave dweller drew in the sand, so it is not surprising that much of our current thinking about GIS databases grew out of traditional cartography.

For millenia, cartographers drew maps with pen and ink, making cartography an art as much as a science. A pen is a vector device; it portrays features as points, lines, or areas circumscribed by lines. Line thickness can be varied somewhat by the use of different pen nibs, but every feature on the map, including lettering, must ultimately be represented by some combination of points and lines. Fortunately, key features that early cartographers needed to portray were also linear: coastlines, streams, roads, property lines, etc.

Map users have become accustomed to vector representations of land features, even when they are abstractions of their true appearance. For example, a stream is often depicted as a single line on a map, although in nature it is an areal expanse of water bounded by land on two sides. Likewise, the use of contour lines is an accepted manner of representing elevation on a map, even though the lines do not exist in reality on land.

The substitution of the computer for pen and ink in map-making has introduced new opportunities for the representation of spatial data. Vectors no longer need be the only means of representation; areal features with imbedded detail, such as an expanse of forest comprising many small clumps of different tree species, can be more realistically represented than was possible with pen and ink in the human hand. Numerical data that had to be generalized in the form of isopleths (e.g. elevational contour lines) can be represented as a continuous data surface. The directionality of streams or other flow paths can be recorded in the database.

The use of computers has, however, presented challenges to those accustomed to using pen and ink for spatial representations, because computers and maps differ fundamentally in the way that they store data. Information on a map is stored graphically in the form of points, vectors, symbols, colors, labels, and reference coordinates. Computers, on the other hand, store data as numbers. Different classes of features are usually represented by numerical codes, which can be combined mathematically with each other or with numerical codes from other data layers. Thus, whereas a map is a visual representation of spatially distributed data, a GIS data layer is a numerical representation. This new means of data storage requires new conventions for data representation.

2.2 Depicting ecological data

Regardless of whether a map is drawn on paper or stored in a computer, it is merely a representation of reality. A digitized map is a computer representation of something that is itself a representation (the source map). Therefore, before working with any GIS database, it is important to think about how the spatial and attribute characteristics of a particular entity may best be represented.

Entities, objects, and features

As used in this chapter, **entities** are real-world phenomena, whereas **objects** are their digital representation. A **feature** is both an entity and its object representation.

2.2.1 Entities as spatial objects

Entities can be represented on a map or data layer as zero-, one-, or two-dimensional spatial objects. A **point** is a zero-dimensional object representing a single location, such as a sample site at the outlet of a lake (Fig. 2.1). A **line** is a one-dimensional sequence of connected points representing a linear feature, such as the lake edge or a stream draining

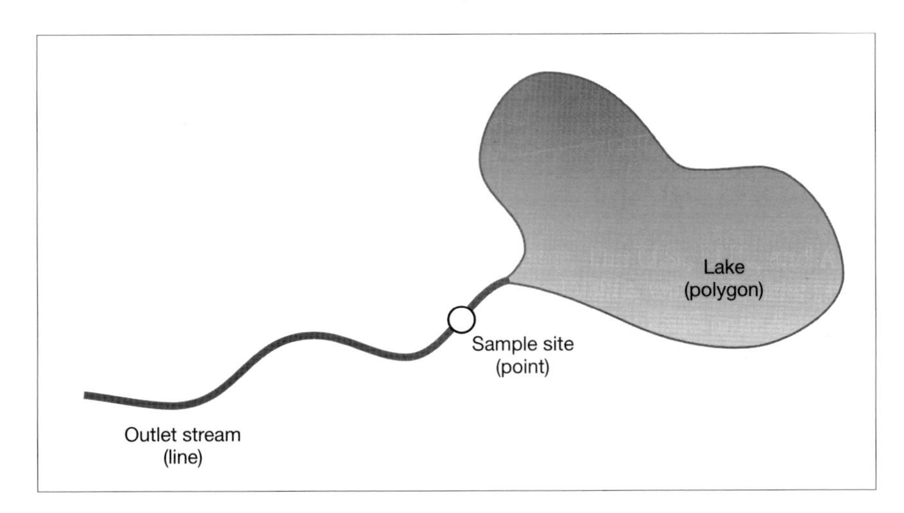

Fig. 2.1 Landscape features representing different spatial primitives.

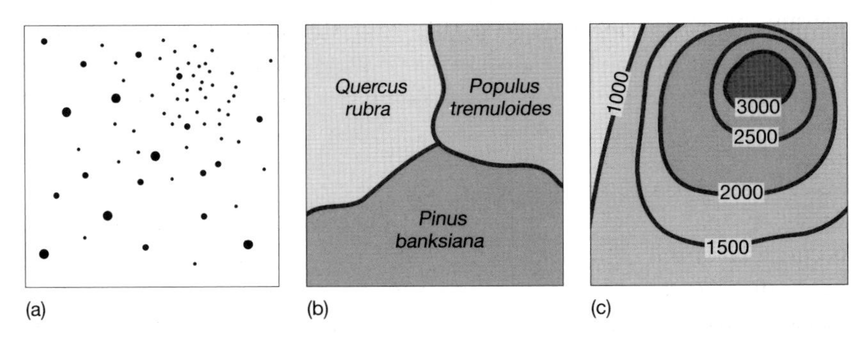

Fig. 2.2 The use of different spatial objects to represent a forest: (a) individual tree stems (points), (b) stands of different tree species (polygons), (c) isolines representing tree stem density per unit area (lines).

the lake. A **polygon** is a two-dimensional area, such as the lake surface. These three types of objects (points, lines, and polygons) are the spatial primitives represented by raster- and vector-based GIS (Peucker & Chrisman 1975). Additionally, some types of GIS can represent three-dimensional **volumes**, such as the water in the lake.

The way in which ecological entities are depicted as spatial objects varies according to the purpose of the analysis and the scale of the representation. A woodlot, for example, may be represented as many point objects (e.g. a stem map), as one or more polygons (e.g. a stand map), or as a series of lines (e.g. an isoline map denoting tree density per unit area: Fig. 2.2). On a small-scale map, the woodlot may be represented as a dimensionless point, even though in reality it has area.

Large scale versus small scale

The terms "large scale" and "small scale" frequently are misused and misunderstood. On a large-scale map, things look large. Buildings would look large on a map of a single city block (a large-scale map), but they would reduce to tiny points on a smaller scale map in which the city block is shown with many others. A large-scale map on a 21 × 29.5 cm page represents a smaller area on the ground than does a small-scale map of the same dimensions.

Cultural features such as buildings and highways often are represented by spatial objects that are two-dimensional miniaturizations of their true appearance, but this representation is rarely the case for biological entities. In fact, individual biological organisms are rarely mapped at all because: (i) they are too small to be detected by most remote sensing techniques; (ii) ground-based mapping of individual organisms is time-consuming and expensive; and (iii) with the exception of immobile, long-lived plants (e.g. trees), the distribution of organisms is spatially and temporally so dynamic that maps of their distribution would become rapidly outdated.

2.2.2 Entity discreteness

An entity that is spatially discrete has discernable limits, such that each occurrence of the entity can be easily distinguished. Discrete entities can be counted (e.g. number of deer, number of stems, number of farm fields). Most anthropogenic features are spatially discrete (buildings, roads, political jurisdictions, property lines, agricultural fields) as are many physical features (bedrock discontinuities, river banks, burn perimeters, the walls of a gopher tunnel).

The boundaries of ecological entities, however, are often indistinct. The boundary between two plant communities, for example, occurs where there is a change in the assemblage of species. In some instances, there may be distinct zonation between the communities (Fig. 2.3a), but more

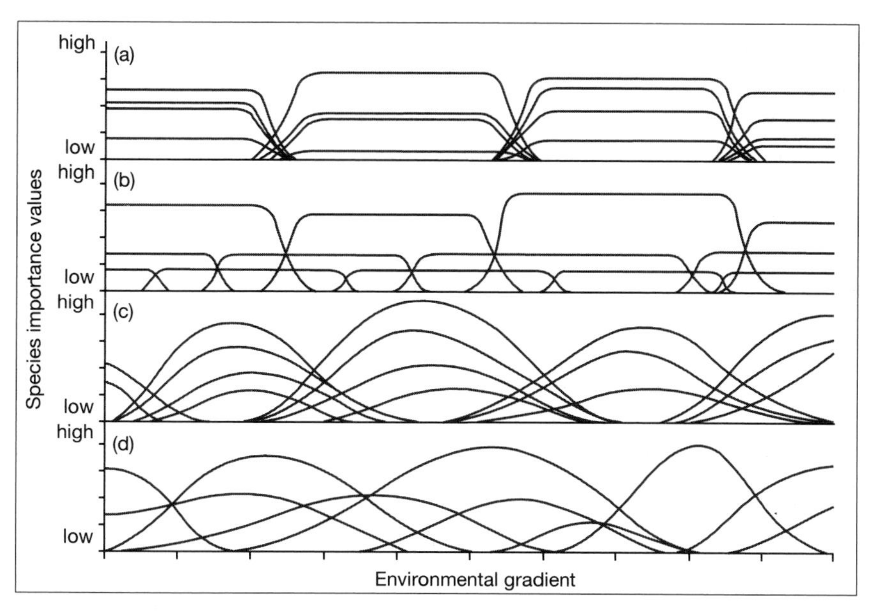

Fig. 2.3 Hypothetical species distributions along an environmental gradient: (a) species occur in associations that have sharp boundaries with one another, (b) competing species exclude one another along sharp boundaries, but do not become organized into groups with parallel distributions, (c) species are organized into groups without sharp boundaries, (d) centers and boundaries of species populations are scattered along the environmental gradient. From *Communities and Ecosystems 2/E*, by Whittaker © 1975. Reprinted by permission of Prentice-Hall, Inc., Upper Saddle River, NJ.

often the change is gradual and does not produce sharp boundaries (Fig. 2.3d).

Boundary discreteness depends on the rate of change with distance, as well as the magnitude of change (Fig. 2.4). The most distinct boundaries are those for which critical ecological properties show a large change over a short distance (Fig. 2.4a). Natural boundaries are more difficult to

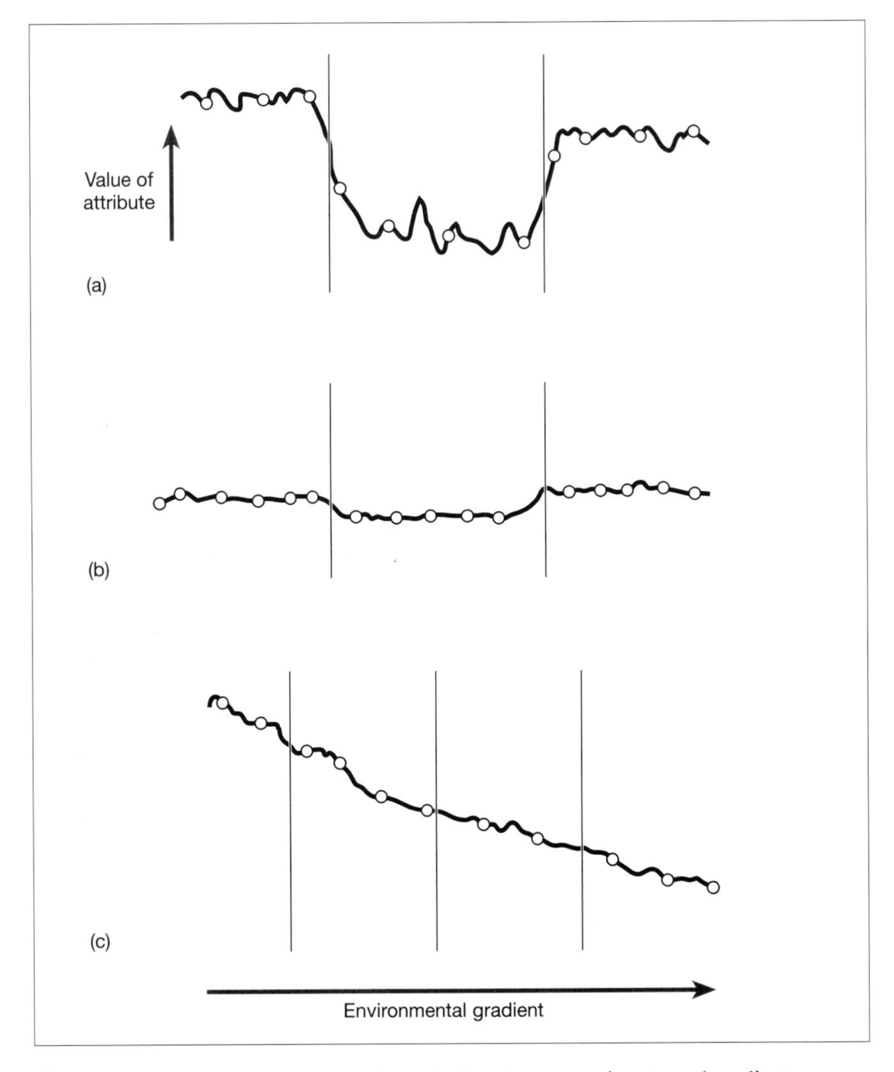

Fig. 2.4 Abruptness and magnitude of boundaries along an environmental gradient, univariate data: (a) abrupt, high-magnitude change, (b) abrupt, low-magnitude change, (c) gradual transition, boundaries arbitrarily placed. Small solid circles represent sample points. After Burrough 1986, by permission of Oxford University Press.

detect when the magnitude of change is abrupt but small (Fig. 2.4b), or gradual but large.

Where ecological properties change along a continuum without natural subdivisions, threshold values may be used to divide the gradient into compartments (Fig. 2.4c). An example is the use of contour intervals to denote constant elevation thresholds. Although there may be statistically significant differences among the individual compartments, the boundaries between them are completely artificial.

Boundaries are almost always depicted as being discrete and one-dimensional (i.e. widthless lines), regardless of their true distinctness and width. Vegetation maps, for example, generally distinguish different categories with distinct boundaries. This form of depiction may be a historical legacy, because early vegetation maps usually were drawn by human cartographers with pen and ink. Such depiction is very appropriate where vegetation boundaries are truly discrete due to anthropogenic manipulation or natural breaks in edaphic conditions, but it is less appropriate where vegetation exists as a continuum. Remote sensing (see Chapter 9) and spatial interpolation of point data (see Chapter 7) provide new algorithms for generating maps of ecological entities that more realistically portray their spatial distribution and discreteness.

2.2.3 Classification of entity attributes

Classification is a means of grouping entities into categories or sets with unifying attributes. People classify natural entities and phenomena in order to organize knowledge, stimulate recognition of group properties, and to elucidate relationships among groups (Buol *et al.* 1980). The way entities are classified in a GIS database greatly influences the scientific outcome of analyses done with those databases, so it is important to understand the basics of classification.

A good classification should meet the following requirements:

- **Must be relevant to the purpose.** The purpose of the classification must be clearly defined, and the classes determined accordingly.
- **Must be comprehensive.** A classification scheme must be capable of categorizing everything within the population of entities being considered. Comprehensiveness is not a function of the number of classes; a three-class scheme may be as comprehensive as a 100-class scheme. Redefining the population to exclude some entities, or adding a "wastebasket" class (e.g. "other") provides comprehensiveness without having to classify unwanted information.
- **Should be based on observable rather than hypothetical properties.** Cline (1963) describes the danger in adopting a classification system based on hypotheses:

A classification system can prejudice the future. If its criteria are

> hypotheses without some device for constant and inescapable scrutiny in relation to fact, the hypotheses become acceptable as fact. Such acceptance can mold research into patterns of the past and can limit understanding of even new experience to concepts based on knowledge of the past.

- **All classes should be discrete from each other.** Even though groups of entities are not always discrete from each other, the classes in a classification system generally should be. In addition to describing the central concept of the class, a class definition should also describe the limits of each class, describe how it is distinguished from similar classes, and provide rules about how difficult class assignments should be resolved.
- **The classification must be repeatable.** Repeatability is best ensured by good documentation.
- **The classification should be applicable to the technology.** A classification should be designed in accordance with the capabilities of the technology used to implement it. For example, Anderson and colleagues (1976) developed a land use/cover classification for use with remote sensors that is still in widespread use. However, the purpose of the classification (1 above) should override technological limitations when the two conflict.

2.2.3.1 Taxonomic classifications versus technical groupings

A classification system with which all biologists are familiar is the system of taxonomy developed by Carolus Linnaeus for organisms. This classification system distinguishes organisms considered to be distinct from each other. The words used to describe each taxonomic class are descriptive, immediately conveying information about the organism (e.g. *Populus tremuloides* is a poplar with trembling leaves). The system is hierarchical, with different levels in the hierarchy indicating the strength of similarity among organism groups. The rules for implementing this classification system are well understood and accepted, but even after two centuries of use there are still variations in how it is implemented. Some taxonomists tend to create new classes for each new variation observed, whereas others tend to group similar organisms into fewer categories. Thus, just as a map is a representation of the spatial distribution of entities, a classification system is a representation of their attributes.

Unlike the Linnean system of classification, in which each new entity receives a new classification, a **technical grouping** has a fixed number of classes into which entities are compartmentalized. Each compartment is:

> . . . a group of individuals tied by bonds of varying strength to a central nucleus. At the center is the modal individual in which the modal properties of the class are typified. In the immediate vicinity

are many individuals held by bonds of similarity so strong that no doubt can exist as to their relationship. At the margins of the groups, however, are many individuals less strongly held by resemblance but more strongly held by similarity to this modal individual than to that of any other class (Cline 1949).

2.2.3.2 Hierarchical versus horizontal classification

Data which have a parent/child or one-to-many relationship may be classified **hierarchically**. In a hierarchical classification system different groups of classes are ranked in some meaningful order, whereas in a horizontal system all classes are on one equal level. Hierarchical organization is best used when a population is so diverse that a horizontal grouping fails to show the desired relationships (Cline 1949). Hierarchical systems should be arranged such that higher categories are classified on the basis of broad, general characteristics which are associated with a number of covarying properties, while lower categories distinguish more specific and detailed qualities. When established in this manner, a hierarchical classification system can provide meaningful information to both generalists and specialists.

An example of hierarchical classification is the U.S. system of wetland classification (Cowardin *et al.* 1979), which contains five systems and ten subsystems, modified by a number of classes, subclasses, water regimes, and other characteristics (Fig. 2.5). A horizontal classification system of this complexity would take hundreds of numbers to identify uniquely each possible combination of wetland descriptors. The hierarchical alternative assigns different codes to the classes at each level of classification, and their combination uniquely identifies the wetland type. GIS databases are conducive to hierarchical classification, because GIS functions can be used to combine or separate data at different levels of hierarchy.

2.2.3.3 Classification in relation to data type

Attribute data are of three basic types: categorical, ordinal, and continuous:

- **Categorical** — classes group entities that are discrete but have no particular sequence or quantitative value (e.g. plant communities, geologic types, soil series). Also called nominal variables.
- **Ordinal** — classes group entities that are discrete and sequential, but not quantitative (e.g. stream orders, birth order).
- **Continuous** — classes represent a continuous range of quantitative data (e.g. temperature, elevation, population density). Not only do the classes have a natural sequence, but the intervals between classes are also meaningful.

When the entities being described have discrete attributes, as is the

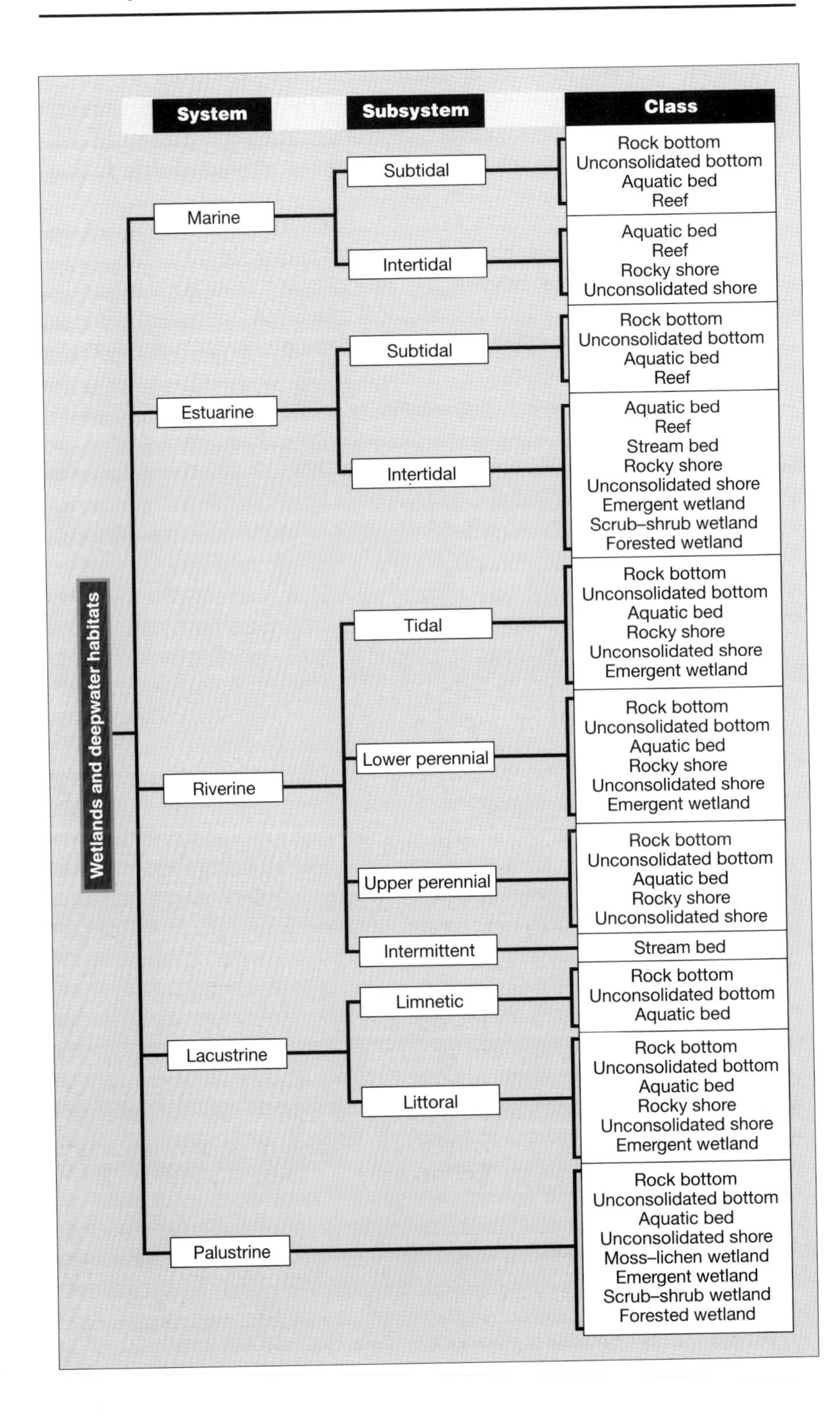
System
Subsystem
Class
Wetlands and deepwater habitats
Marine
Subtidal
Rock bottom
Unconsolidated bottom
Aquatic bed
Reef
Intertidal
Aquatic bed
Reef
Rocky shore
Unconsolidated shore
Estuarine
Subtidal
Rock bottom
Unconsolidated bottom
Aquatic bed
Reef
Intertidal
Aquatic bed
Reef
Stream bed
Rocky shore
Unconsolidated shore
Emergent wetland
Scrub–shrub wetland
Forested wetland
Riverine
Tidal
Rock bottom
Unconsolidated bottom
Aquatic bed
Rocky shore
Unconsolidated shore
Emergent wetland
Lower perennial
Rock bottom
Unconsolidated bottom
Aquatic bed
Rocky shore
Unconsolidated shore
Emergent wetland
Upper perennial
Rock bottom
Unconsolidated bottom
Aquatic bed
Rocky shore
Unconsolidated shore
Intermittent
Stream bed
Lacustrine
Limnetic
Rock bottom
Unconsolidated bottom
Aquatic bed
Littoral
Rock bottom
Unconsolidated bottom
Aquatic bed
Rocky shore
Unconsolidated shore
Emergent wetland
Palustrine
Rock bottom
Unconsolidated bottom
Aquatic bed
Unconsolidated shore
Moss–lichen wetland
Emergent wetland
Scrub–shrub wetland
Forested wetland

Table 2.1 Methods for selecting class intervals for univariate continuous data. After Evans 1977.

Method	Description	Examples
Arbitrary	Intervals chosen without any clear aim, often irregularly spaced	–
Exogenous	Threshold levels are relevant to, but not derived from, the data set under study	Standard slope classes used by the U.S. Soil Survey (Soil Survey Staff 1993)
Idiographic	Intervals chosen with respect to specific aspects of the data set	Multimodal groupings derived from histograms of data frequency
Serial	Interval limits are mathematically related to each another	Percentiles dividing the data into classes of equal frequency Classes centered on the mean and defined as a proportion ± the standard deviation Equal intervals on arithmetic scales Equal intervals on reciprocal, trigonometric, geometric, or logarithmic scales

case for categorical and ordinal data, defining class limits is fairly straightforward. Rules can be developed that assign an entity to one class or another, such as deciduous versus coniferous trees. Of course, problems can arise when an entity has characteristics of both classes (e.g. a larch [*Larix laricina*] tree is both deciduous and coniferous), but these can usually be avoided by careful use of terminology and definition of class limits. Written documentation describing the classes, their derivation, and their limits is an essential but often overlooked component of scientific database management.

Defining class limits for continuous data requires a more analytical approach, and has been the subject of much research. Methods for selecting class intervals for univariate continuous data have themselves been classified by Evans (1977; Table 2.1). Of these, serial classes with equal arithmetic intervals, such as contour intervals for elevation, is one of the most common schemes used. Tomlin (1990) additionally divides serial classes into those with **interval** and **ratio** scales. An interval scale does not have an absolute zero or starting point, whereas a ratio scale does. Two scales commonly used to classify temperature, Celsius and Kelvin, illustrate this difference. The Kelvin scale (a ratio scale) has an absolute zero, equal to the temperature at which substances possess minimal energy, whereas the Celsius scale (an interval scale) has a zero

Fig. 2.5 (*Opposite*) Classification hierarchy of wetlands and deepwater habitats used in the National Wetlands Inventory being conducted by the U.S. Fish and Wildlife Service. Subsystem categories are distinguished by water depth or tidal influence; class categories are distinguished by vegetation or substrate type. From Cowardin *et al.* 1979.

Table 2.2 A simple relational database.

Row	Species code	Species name
1	23	*Acer rubrum*
2	5	*Taxodium distichum*
3	62	*Magnolia virginiana*
4	9	*Pinus taeda*

point arbitrarily defined as the freezing point of pure water. The divisions of a ratio scale are multiplicative (i.e. a 20 K object is twice as warm as a 10 K object), whereas those of an interval scale are merely additive (i.e. a 20°C object is 10°C warmer than a 10°C object, but it is not twice as warm as that object). Most GIS contain analytical capabilities for developing classification schemes from continuous data sets, so the use of raw data, rather than pre-classified data, is preferable when the derivation and accuracy of a classification are unknown.

The above examples apply to **univariate** continuous data, in which only one attribute is being measured at each data point. However, in many environmental studies, multiple attributes are known for each data point. Techniques for classifying these **multivariate** data include cluster analysis and ordination. Multivariate techniques are commonly used in plant ecology and remote sensing image analysis, and detailed descriptions of their application can be found in the literature of those fields (e.g. Pielou 1984; Lillesand & Kiefer 1994).

2.2.4 Using numbers to represent entity classes

Most GIS require the use of number codes for entity classes, whether the data those numbers represent are categorical, ordinal, or continuous. The use of numbers is intuitive for ordinal and continuous data, but requires development of a **relational database** for categorical data, so that the numbers can be related to the entities they represent. A simple relational database for selected tree species is shown in Table 2.2. Each row in the database represents a single data record, and each column represents an

Table 2.3 A relational database with multiple attributes.

Row	Species code	Species name	Common name	Leaf form	Leaf retention
1	23	*Acer rubrum*	Red maple	Broad-leaved	Deciduous
2	5	*Taxodium distichum*	Bald cypress	Needle-leaved	Deciduous
3	62	*Magnolia virginiana*	Sweet bay	Broad-leaved	Evergreen
4	9	*Pinus taeda*	Loblolly pine	Needle-leaved	Evergreen

Table 2.4 A relational database for spatial data.

Row	Polygon number	Species code
1	0	0
2	1	62
3	2	9
4	3	23
5	4	9
6	5	5
7	6	5
8	7	23

attribute, either its numeric code or its species name. Each of these tree species can be categorized in other ways, such as common name, leaf form, and leaf retention (Table 2.3). The categories describe different tree characteristics, but all have a direct relationship to the entities represented by the codes.

Many GIS have a structure in which spatial data are stored in one relational database, attribute data are stored in a second relational database, and the two are linked by a common set of codes referring to the features they represent. A relational database for a stand map of the above trees is shown in Table 2.4. In this case, the polygon codes represent individual stands, with "0" representing those areas that are not forested. The species codes common to this table and the attribute table (Table 2.3) allow the two data sets to be linked.

GIS software that uses 8-bit notation limits the number of classes to integers between 0 and 255 (see Section 1.4.1). Real (decimal) numbers can be converted to fit within this range by the use of a **scalar**. For example, for data set values ranging between 0.1 and 300.1, the scalar can be computed as:

$$\frac{255}{(300.1 - 0.1)} = 0.85$$

Each value in the database is multiplied by the scalar (0.85), and the value of that product rounded to the nearest integer and assigned to the new database.

2.2.5 Spatial resolution

Because a GIS makes it easy to change the scale at which ecological features are displayed, many GIS users overlook the spatial resolution, or detail, of the source data. As used in photography, the resolving power of a film is primarily a function of the size distribution of the silver halide grains in the emulsion (Lillesand & Kiefer 1994). As used in conventional mapping, spatial resolution is measured in terms of the **minimum mapping unit** (MMU), the smallest feature that can be uniquely represented.

In some cases, the MMU may be a function of anthropogenic thresholds, such as the minimum area of wetland regulated or the minimum parcel size in a suburban housing development. In other cases, the MMU may be a natural threshold, such as the minimum area of glacial kames or kettleholes.

The MMU varies with the type of feature, even within the same map. For example, vector land use/land cover (LULC) maps produced by the U.S. Geological Survey (USGS) have a minimum mapping unit of 4 ha for urban features and water bodies, but a minimum mapping unit of 16 ha for rangeland, forest, wetlands, tundra, snowfields, most agricultural lands, and most barren lands. Similarly, the U.S. National Wetlands Inventory routinely maps wetlands smaller than 0.5 ha in the prairie pothole region, but generally uses a larger MMU for forested wetlands (Tiner 1990). This difference in mapping resolution is due to the differences in contrast between these features and their surroundings on the aerial photographs that are the data source for these maps; the contrast between a small water body and a prairie is much greater than the contrast between wetland and non-wetland forest. The rectilinearity of urban features also contrasts with the rounded lines of natural features, making them easy to distinguish from each other.

Different MMUs within the same map also result when some resources are more valued than others. For example, a forest photograph interpreter would map stands of merchantable timber in more detail than stands of non-merchantable lowland brush for a map of forest economic potential.

The resolution of the source materials and tools used to make a map also influences the MMU achievable. The MMU definable from a satellite image is generally equal to the area of four adjacent pixels. The scale of aerial photography used also limits the MMU; smaller features can be detected and delineated on larger scale photography. When a pen is used to delineate features directly on an aerial photograph, the MMU is also affected by the size of the pen point. Tobler (1988) observed that the smallest physical mark that can be made by a cartographer is about 0.5 mm in size, and used that observation to develop a rule of thumb for calculating resolution (i.e. the smallest mark that can be made) and detection limits: divide the denominator of the map scale by 1000 to get the detectable size in meters, and by 2000 to get the resolution. For example, the resolution of a 1 : 24 000 photograph would be 12 m (i.e. the width on the ground represented by a 0.5-mm pen line) and its minimum detectable size would be 24 m.

Expressing map scale

Map scale is expressed graphically with a scale bar, or textually with a **representative fraction**. A representative fraction is the ratio between a map distance and the distance it represents on the ground. Both the numerator and denominator use the same units, so the fraction is dimensionless. For example, at a scale of 1 : 10 000, 1 cm on the map represents 10 000 cm (100 m) on ground. Expressing map scale as inches-to-miles is also commonly used, but less desirable because it does not use the International System of Units (SI).

When not stated in the documentation for the data source, the MMU can be determined with a cumulative frequency curve or frequency histogram of the patch sizes in a particular class. In the example shown in Fig. 2.6 of beaver impoundment areas, the MMU is also the modal impoundment area (= 1 ha). Notice that a MMU of 14 ha would be too large to detect most of these beaver impoundments, because 95% of them are smaller than 14 ha.

Hierarchy theory (Allen & Starr 1982) defines the scale of a structure as "the time and space constants whereby it receives and transmits information," a concept that is useful in the context of GIS databases (Aspinall & Pearson 1996). Spatial patterns that are mapped at too fine a resolution impart excessive information (i.e. "noise"), whereas those mapped at too coarse a resolution appear as constants (Shugart *et al.* 1991). The most suitable resolution is one that conveys meaningful information about a feature of interest without excessive noise.

2.2.6 Data quality

Many GIS users assume that maps are accurate renditions of reality, and do not investigate or question their derivation. However, the adage "gargage in, garbage out" is as applicable to GIS analysis as it is for any computer analysis. Just as an understanding of the methods and source data used in a scientific experiment is crucial to evaluating results, the methods and source information used to construct a map are crucial to evaluating the results of a GIS analysis.

Metadata, descriptive information about a data set, are important to assessing data quality (Chrisman 1994; Lanter 1994). GIS metadata are often conveyed in the form of a **data dictionary**, which conveys the meaning and structure of entity and attribute data (National Institute of Standards and Technology 1992). A data dictionary specifies the type and range of values each attribute may take, defines the meaning of attribute value codes, gives units of measurement (if appropriate), defines the

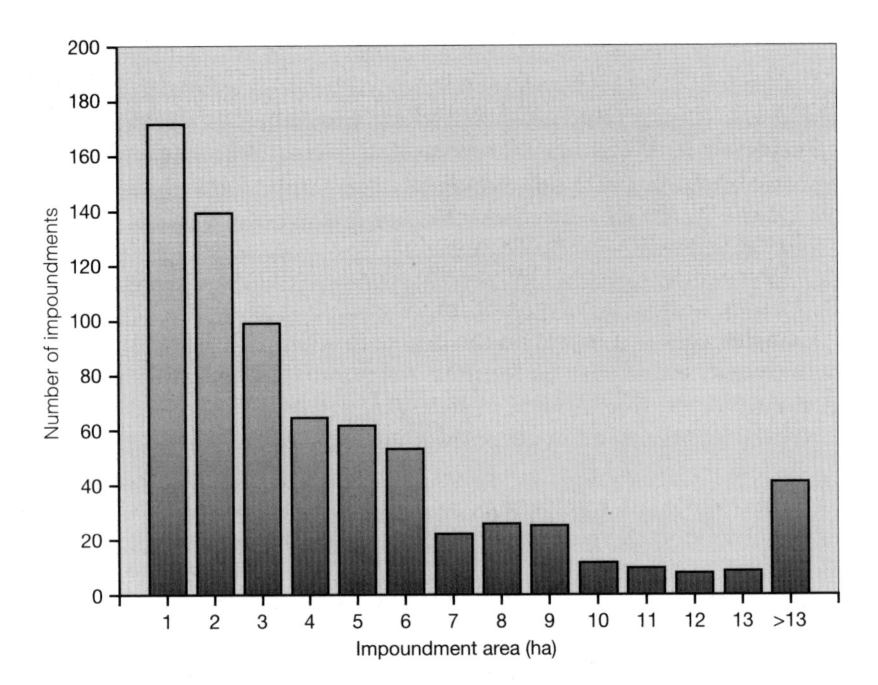

Fig. 2.6 Size frequency distribution for area of beaver impoundments at Voyageurs National Park. From Johnston, C.A., unpublished data.

meaning of entity and attribute labels (labels printed on hard copy output are usually textual, whereas attribute codes stored within a GIS are usually numerical; see Section 2.2.4), identifies an authority for each definition, and describes the layout and format of records in the attribute database.

The issue of data quality in GIS has received extensive review (Walsh *et al.* 1987; Goodchild & Gopal 1989; Thapa & Bossler 1992; Goodchild 1993; Aspinall & Pearson 1996; Burrough *et al.* 1996) and GIS data standards are being developed in several countries. The U.S. National Committee for Digital Cartographic Data Standards has identified five major components of data quality (Davis *et al.* 1992; FGDC 1994):

- **Lineage** — a description of the source materials (identification and vintage), methodologies, and transformations used to build the database. Methodological descriptions should be provided for the acquisition and compilation of the source data (e.g. a source map) as well as for the entry and transformation of those data in the GIS.
- **Positional accuracy** — the accuracy of the coordinates within the database.
- **Attribute accuracy** — the accuracy of attribute identification, classification, and quantification. Burrough *et al.* (1996) suggest several factors

that can affect the quality of attribute data: methods of measurement, recording and analysis techniques, assumptions about the kind of spatial/temporal variation (discrete or continuous), spatial and temporal resolution, spatial and temporal variability, number or density of observations, data interpolation methods, and numerical representation in the computer (integer/real/double precision).

- **Logical consistency** — a report on the occurrence of problems, such as lines intersecting at points other than nodes, lines on top of other lines, lines failing to close a polygon of importance to the database, etc.
- **Completeness** — a report on the degree to which the database exhausts the universal set of various features; for example, does a transportation database include all roads or just federal highways?

As the influence of error propagation has been increasingly recognized, new tools have been developed for its detection and quantification (Veregin 1994). An error propagation tool called ADAM, which uses statistical theory of error propagation and model approximation techniques to calculate errors within a GIS, has been developed by Heuvelink (1993; c.f. Burrough *et al.* 1996). ADAM analyzes a GIS model, recommends an error propagation strategy (Taylor series approximation, Rosenbleuth's method, or Monte Carlo conditional simulation), and compiles a control file which computes the results using standard GIS commands.

2.3 GIS data structures

Although all GIS store data about the location and attributes of real-world entities, some use **raster** data structures and some use **vector** data structures. The digital representation of spatial data is a complex and evolving field, and the following discussion is only rudimentary. For more information, consult Peuquet (1984), Burrough (1986), and Morrison and Wortman (1992).

2.3.1 Raster data structures

A raster database portrays features as a matrix of equal-area cells that are usually square (Fig. 2.7). The smallest non-divisible element in a raster database is a **grid cell**; the smallest non-divisible element in a digital image is a **pixel** (picture element). Grid cells and pixels are inherently two-dimensional, even though the entities they represent may be zero- or one-dimensional. Therefore, features that are dimensionless (e.g. points, boundaries) or smaller than the minimum raster dimension (e.g. wells, streams) are difficult to depict in a raster GIS. At coarse levels of resolution (i.e. large rasters), polygons appear blocky and lines appear stair-stepped. At finer levels of resolution, a raster representation looks more like a map, but data storage requirements increase exponentially.

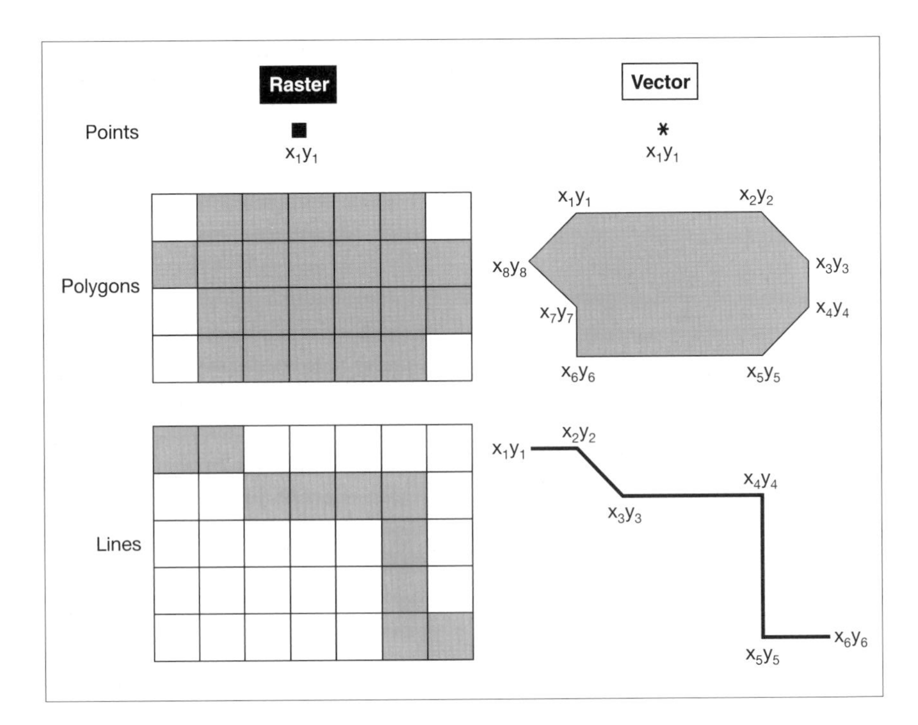

Fig. 2.7 Raster and vector representation of spatial primitives.

While finer resolution raster databases are aesthetically more appealing, the increased detail which they provide may be unnecessary for data analysis. The appropriate raster size should be comparable to the scale at which the ecological process of interest is operating. For example, global circulation models commonly use cells covering degrees of latitude and longitude, sub-continental analyses commonly use square kilometer cells, and regional analyses commonly use 30 × 30 m cells (the size of a Landsat Thematic Mapper pixel). Rasters which are too fine can obscure features at coarser scales by providing too much detail.

Each cell in a raster database is assigned one, and only one, number. That number may represent an individual attribute (e.g. tree species, stem density) or it may represent a spatial object (e.g. tree stands, Table 2.2) that can be linked to multiple attributes via a relational database (e.g. stand attributes, Table 2.3). In addition to numbers representing information internal to the map, a null value must be assigned to portions of the grid array that are outside of the mapped area, comparable to the margins of a paper map that do not contain spatial data.

Quadtrees are a type of raster data storage structure which take into account the inherent resolution of the data. The quadtree is a hierarchical raster data structure based on the successive division of a $2^n \times 2^n$ array into quadrants, in which the data for four quadrants having the same value are aggregated into the next largest block in the hierarchy. As a result, coarse-grained features are stored in large cells, while fine-grained features are stored in small cells.

2.3.2 Vector data structures

A vector database portrays features as points, lines, or polygons (Fig. 2.7). Homogeneous patches are bounded by lines, as they would appear on a conventional map. Because of this similarity, vector-based GIS are preferred for automated cartography.

Most commercially available vector GIS use structures which associate **topology** (i.e. the connectivity or contiguity of geographic features) with spatial objects representing points, lines, and polygons. The terminology for these spatial objects has varied, but was standardized recently in the U.S. by the adoption of the Spatial Data Transfer Standard (SDTS 1992), described in Federal Information Processing Standard (FIPS) Publication 173 (National Institute of Standards and Technology 1992). The SDTS borrows terms from object-oriented computer science to build a general concept of a spatial database. It allows vector and raster data of many different types, models, and structures to be exchanged between dissimilar systems. The SDTS addresses issues ranging from conceptual modeling to details of physical encoding and definitions of feature and attribute terms (Fegeas *et al.* 1992).

The SDTS defines a set of 13 basic zero-, one-, and two-dimensional objects, two of which are raster data structures (pixel, grid cell) and the remainder of which are vector (Tables 2.5–2.7; Fig. 2.8). The vector structures are of two types: (i) geometry-only spatial objects; and (ii) geometry and topology spatial objects. For example, a **point** is a one-dimensional geometry-only object that specifies geometric location,

Table 2.5 Terminology used in the SDTS for zero-dimensional spatial objects.

Geometry-only spatial objects	
Point	An object that specifies a single geometric location
Point subtypes:	
Entity point	A point used to identify the location of a feature such as a building, tower, buoy, etc.
Label point	A point used to identify the location of text or symbology on a display
Area point	A point representing an area for purposes of storing attribute information about the area
Geometry and topology spatial objects	
Node	An object that represents the junction of two or more links or chains, or the termination of a one-dimensional object

Table 2.6 Terminology used in the SDTS for one-dimensional spatial objects.

Geometry-only spatial objects	
Line segment	An object representing a straight line connecting two points
String	An ordered sequence of connected, non-branching line segments (a string may intersect itself or other strings)
Arc	A curve that is defined by a mathematical function
G-ring	A sequence of non-intersecting strings and/or arcs that close to form the boundary of an area. A ring represents a closed boundary, but not the interior area inside the closed boundary
Geometry and topology spatial objects	
Link	A topological connection between two nodes
Chain	A directed, non-branching sequence of non-intersecting line segments and/or arcs bounded by nodes
Chain subtypes:	
Complete chain	A chain that explicitly references left and right polygons and beginning and ending nodes
Area chain	A chain that references left and right polygons, but not beginning and ending nodes
Network chain	A chain that references beginning and ending nodes, but not left and right polygons
GT-ring	A ring created from complete and/or area chains. A ring represents a closed boundary, but not the interior area inside the closed boundary

Table 2.7 Terminology used in the SDTS for two-dimensional spatial objects.

Geometry-only spatial objects	
Interior area	An area not including its boundary
G-polygon	An area consisting of an interior area, one outer G-ring and zero or more non-intersecting, non-nested inner G-rings
Pixel	A two-dimensional picture element that is the smaller non-divisible element of a digital image
Grid cell	A two-dimensional object that represents the smallest non-divisible of a grid
Geometry and topology spatial objects	
GT-polygon	An area that is an atomic two-dimensional component of one and only one two-dimensional manifold. A GT-polygon can be associated with its chains
GT-polygon subtypes:	
Universe polygon	The part of the universe outside the perimeter of the area covered by the other GT-polygons
Void polygon	Defines a part of the two-dimensional manifold that is bounded by other GT-polygons, but otherwise has the same characteristics as the universe polygon

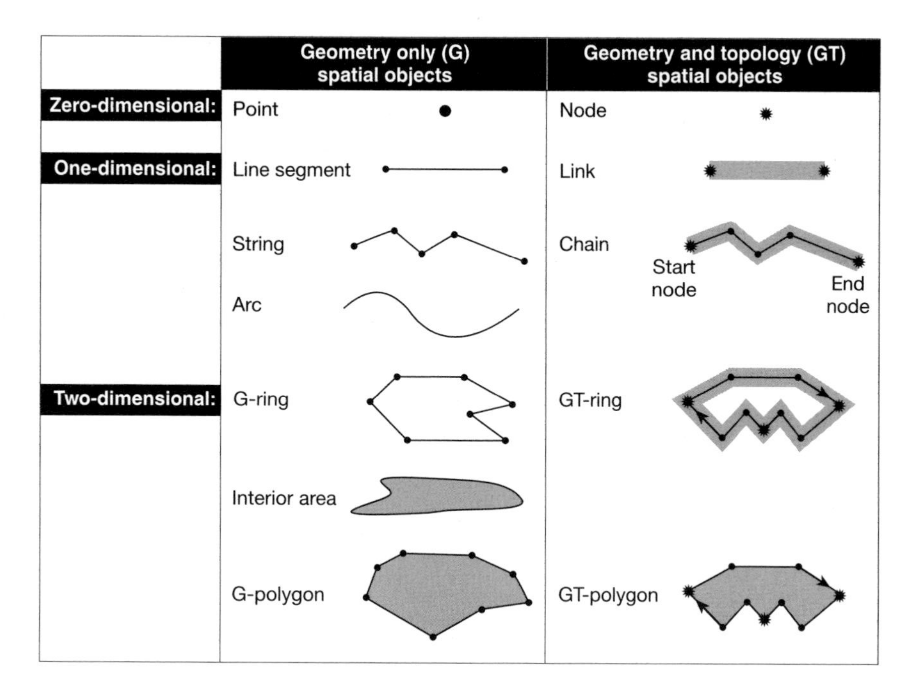

Fig. 2.8 Vector spatial objects in the Spatial Data Transfer Standard.

whereas a **node** is a point that is a topological junction of two or more links or chains. The topology of a node identifies the line segments for which it is an endpoint. A **string** is a geometry-only ordered sequence of connected, non-branching line segments that may or may not intersect itself or other strings, whereas a **chain** is a directed string bounded by nodes. Topology identifies the starting and ending points of a chain.

Like a raster GIS, a vector GIS also assigns attributes to spatial features. For example, a point database of tree stems may be associated with data about tree species, stem diameter, or tree height. Attribute data may be stored in the same file with the spatial data or they may be kept in a separate table as a relational database.

2.3.3 Raster versus vector data structures

There has been much debate among GIS users about the relative merits of raster versus vector data structures. However, these distinctions become blurred as raster and vector GIS increasingly share analysis and conversion capabilities. For scientific applications, the choice of data structures should be based primarily on: (i) the discreteness of the entity being depicted; (ii) the intended application; (iii) the source data. In general, a vector structure is preferable if the entities being depicted are inherently discrete and must be represented precisely (e.g. roads, ownership boundaries), or if the intended analysis involves measurement or analysis of linear features (e.g. stream networks). A raster structure is generally preferable when the entities being depicted have indistinct boundaries, when the data are continuous (e.g. elevation), when the intended analysis involves spatially distributed modeling (see Chapter 10), or when the data are already in raster form (e.g. an image, a raster GIS database).

2.4 Georeferencing

A distinguishing feature of a GIS is that all data are **georeferenced** to a coordinate system. In a planar **Cartesian coordinate system**, the location of a point is its distance from two intersecting, usually perpendicular, straight lines, the distance from each line being measured along a straight line parallel to the other. Coordinates are usually expressed as *x,y* pairs, as in a raster database. The Universal Transverse Mercator coordinate system (**UTM**, explained in Section 2.5 below) is a planar Cartesian Earth coordinate system, as are **State Plane Coordinate Systems**, used in the U.S. for georeferencing individual states.

A planar coordinate system would work perfectly well for a flat Earth, but as we have known since Christopher Columbus' 1492 voyage (and before), the surface of the Earth approximates a sphere. Thus, although planar Cartesian coordinate systems work well for small areas of the Earth's surface, no two-dimensional Cartesian system would suffice for the globe (however, three-dimensional Cartesian coordinates are used by Global Positioning Systems [GPS] to define Earth locations in space – see Chapter 8). The system of Earth coordinates developed to describe spherical locations is **latitude–longitude**. Lines of longitude are the set of circles that intersect both the north and south pole. They are measured in degrees east or west of the prime meridian that passes through Greenwich, England, which is the origin (0° longitude), with longitude 180° being halfway around the world from Greenwich. Lines of latitude are a series of circles concentric to the north and south poles, measured in degrees north and south of the equator (0° latitude), with the poles at 90° latitude. Lines of latitude are parallel (hence also referred to as parallels) and equidistant, but because lines of longitude converge at the poles, the distance between them varies with latitude. This trait makes latitude–longitude a difficult coordinate system to use on the ground without special equipment.

The Earth is not a true sphere, so a model of the Earth is needed to accurately plot the latitude, longitude, and elevation of features on its surface. In 1687, Sir Isaac Newton first postulated that the Earth was slightly ellipsoidal in shape rather than spherical as previously assumed. With the advent of modern satellites, we now know that the Earth is slightly pear-shaped, with a slight hump around the North Pole, a slight depression around the South Pole, and a slight bulge just south of the equator (Langley 1992). The challenge is to accurately portray the size and shape of the Earth in a way that is useful to cartographers and surveyors.

A **geoid** model attempts to represent the surface of the entire Earth over both land and ocean as though the surface resulted from gravity alone (Dana 1995). The geoid is a representation of the surface of the sea as if extended through the land, and undulates with the effects of gravity (Cheves 1997). Over the years scientists have developed more accurate mathematical models to represent the geoid as techniques and instruments have improved. The World Geodetic System 1984 (WGS-84) geoid defines geoid heights for the entire Earth; the U.S. National Geodetic Survey Geoid-96 is the most recent geoid model for the U.S.

Geographic **datums** and the coordinate reference systems based on them were developed to describe geographic positions for surveying, mapping, and navigation (Dana 1995). Datum types include horizontal (e.g. NAD-27; NAD-83), vertical (e.g. NAVD-88), and complete datums. Different nations and agencies use different datums as the basis for coordinate systems used to identify positions in GIS. Coordinate values resulting from interpreting latitude, longitude, and height values based on one datum as though they were based on another datum can cause positional errors of up to 1.5 km (Fig. 2.9), so it is essential to know which datum was used to georeference a map.

Georeferencing has been revolutionized by the development of GPS (see Chapter 8) that use computer and satellite technology to measure ground location.

2.5 Map projections

Cartographers have long been challenged with the problem of representing the Earth's spherical form on a planar surface, known as **map projection**. There is a bewildering array of map projections, each with different properties with regard to the preservation of shape, area, distance, and direction. A projection can be envisioned as if a light source in the middle of the Earth were casting shadows onto a giant piece of paper wrapped in a cylinder or cone.

A commonly used cylindrical projection is the Transverse Mercator, also known as the Gauss–Krüger projection, in which the cylinder is wrapped around both poles, tangential to the Earth along a line of longitude (Fig. 2.10). A variant, the Universal Transverse Mercator (UTM) projection, uses a secant cylinder (i.e. one that slices a thin ring from the Earth's surface) that has a diameter slightly less than that of the Earth, such that its intersection with the Earth's surface consists of two

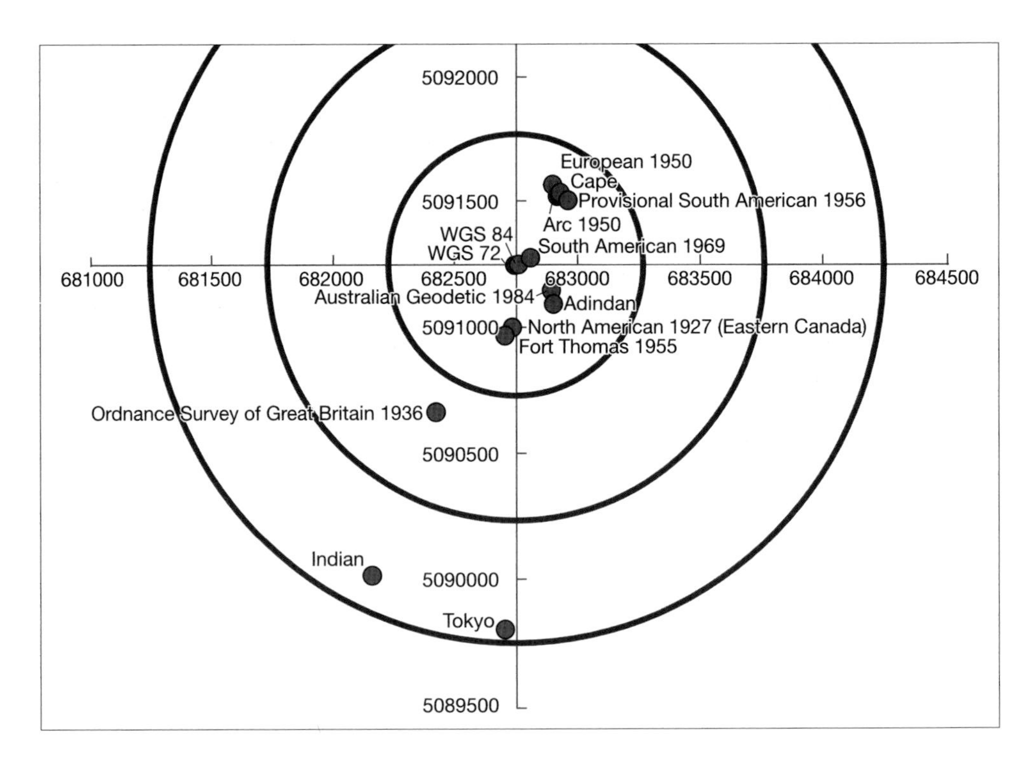

Fig. 2.9 Effect of different datums on UTM eastings and northings (m) of a point with WGS-84 geographic coordinates of 45° 57′ 0.96′′N, 66° 38′ 32.22′′ W. From Featherstone & Langley 1997.

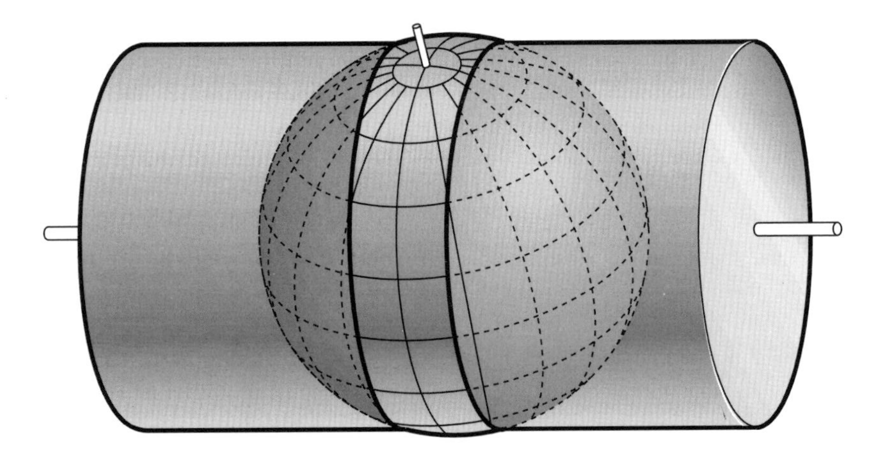

Fig. 2.10 Transverse Mercator supposition of cylinder secant to globe of derivation. From Greenhood 1964. Reprinted by permission from *Mapping* by Greenhood, © 1964 by the University of Chicago.

circles parallel to and 3° on either side of its central meridian. These 6° zones extend from 80°N to 80°S latitude (Fig. 2.11). This projection is **conformal**, preserving the shapes of mapped features. The UTM coordinate system developed for this projection uses Cartesian coordinates, such that each grid square in every 6° wide zone is of the same shape and size. Easting and northing distances are measured in meters relative to the origin of the coordinate system, which lies at the intersection of the Equator and the central meridian of each zone. This is a popular coordinate system for large-scale maps because of its ease of use, but is inappropriate for maps covering an area much wider than the 6° zones.

A commonly used conical projection is the **Lambert conformal conic**, in which features are projected onto a cone secant to the Earth at two latitudes chosen as the standards, and parallels are spaced at increasing intervals with distance north or south of the standard parallels. Maps of the U.S. at 1 : 2 500 000 or smaller published by the USGS use a Lambert

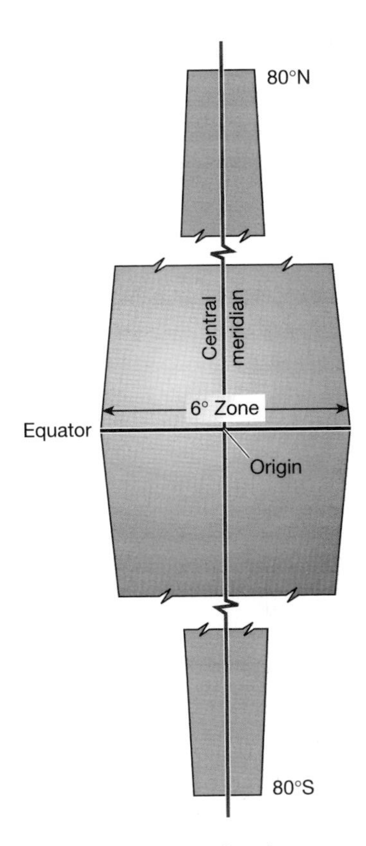

Fig. 2.11 Diagrammatic representation of a gore forming one zone of the Universal Transverse Mercator projection. From Maling 1989, by permission of Butterworth-Heineman Publishing.

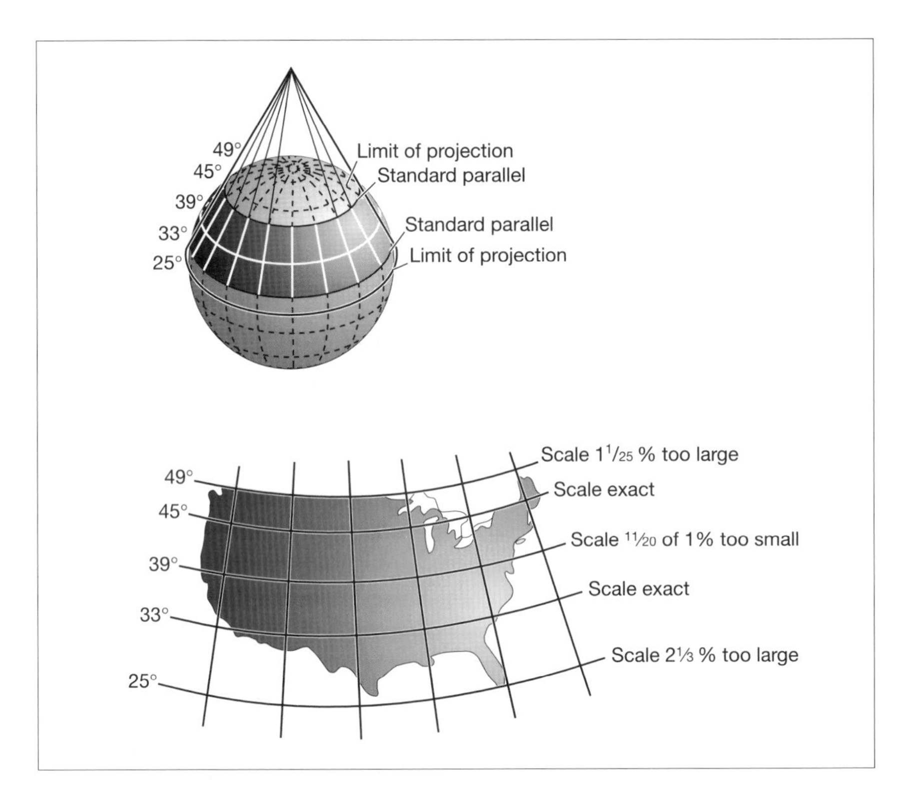

Fig. 2.12 Lambert Conic Conformal projection and resultant grid. Reprinted by permission from *Mapping* by Greenhood, © 1964 by The University of Chicago.

conformal conic projection with standard parallels at latitudes 33°N and 45°N (Fig. 2.12). The International Map of the World also uses a Lambert conformal conic projection, with each map sheet having a dimension of 4° latitude and 6° longitude, and standard parallels at 1/6 and 5/6 of the north–south extent of the sheet. Another commonly used conical projection is **Alber's Equal-Area**. Like the Lambert conformal conic, the Alber's Equal-Area is projected on a cone secant to the Earth. However, its standard parallels are located at 29½°N and 45½°N, and parallels are spaced at decreasing intervals with distance north or south of the standard parallels. As conveyed by its name, this equal-area projection portrays features in proportion to the areas they represent on the ground. A third commonly used conical projection is the **polyconic**, which uses several nested cones tangential to lines of latitude.

2.6 Data sources

Given the time and expense required to develop a GIS database, many GIS users prefer to use existing databases. Many databases are available at continental to global scales, but fewer are available at the ecosystem to landscape scales at which many ecologists work. However, several countries are beginning to generate GIS databases at scales of 1 : 25 000 and larger. Published references provide information on some available data sets (e.g. FGDC 1992; Steyaert 1996), but the most up-to-date source of information on databases is the World Wide Web (WWW: see Section 2.6.3.1 and Appendix).

2.6.1 Major global and continental data sources

UNEP. The United Nations Environmental Programme (UNEP) has been involved in the acquisition and dissemination of environmental data since the founding of the Global Environmental Monitoring System (GEMS) in 1974 (Mooneyhan 1988). The Global Resource Information Database (GRID) was established to provide georeferenced environmental information and access to a unique international data service to help address environmental issues at global, regional, and national levels. GRID data holdings include global data sets for political and natural boundaries, elevation, soils and soil degradation, vegetation, weekly vegetation index, human population, cultivation intensity, ecosystems, lifezones, wetlands, precipitation and temperature anomalies, temperature and moisture availability surfaces, and ozone distribution. GRID also maintains databases for Africa and a number of countries in Africa, Latin America, and Asia. The GRID Programme is centered in Kenya, with regional centers in Thailand, Switzerland, Nepal, Japan, Poland, Norway, Brazil, and the U.S.

Digital Chart of the World. The Digital Chart of the World (DCW) is a 1 : 1 000 000-scale base map of the world developed by the U.S. Defense Mapping Agency (DMA) with the cooperation of Australia, Canada, and the U.K. (DMA 1992). The primary source of the database is the DMA Operational Navigation Chart series, which is the only civilian map series having consistent, continuous global coverage of essential map features (Steyaert 1996). The DCW is organized into 17 thematic layers including political boundaries, ocean coastlines, cities, transportation networks, drainage, land cover, and elevation contours. The database also includes an index of more than 100 000 place names worldwide. For some regions of the world, data such as elevation must be used with caution because of errors in the source maps, and should be checked for accuracy and consistency (Steyaert 1996). Ordering information is in the Appendix.

Elevation. Digital elevation models (DEMs) provide raster format topographic data essential for a variety of ecological analyses (see Chapter 4). The availability of consistent DEM data for the globe is currently limited to ETOPO5 (nominally 10×10 km MMU). Work is in progress at the USGS EROS Data Center to develop a consistent global DEM at an approximate resolution of 1 km^2 by interpolating digital elevation contours with the DCW (Moore *et al.* 1993). ETOPO5 can be retrieved via anonymous ftp over the Internet (see Section 2.6.3.2 and Appendix).

Land cover. The EROS Data Center has compiled 1 km^2 resolution land cover data from Advanced Very High Resolution Radiometer (AVHRR) satellite data (see Chapter 9) for the conterminous U.S., Alaska, Mexico, and Eurasia (Eidenshink 1992; Brown *et al.* 1993; Loveland & Scholz 1993). Annual data sets (1990 to present) for the conterminous U.S. contain 159 land cover classes, each representing unique combinations of land cover mosaics, seasonal properties (onset and peak of greenness, and duration of green period), and relative levels of primary production. Available on CD-ROM for a nominal fee, the data consist of: (i) the AVHRR data used in the analysis; (ii) the initial and final interpretations of the land cover regions; (iii) descriptive and quantitative attributes of the land characteristics for each of the land cover regions; and (iv) derived thematic interpretations of the land cover regions (for example, aggregation to the USGS's land cover scheme, and biophysical interpretations such as the onset of greenness).

The U.S. Environmental Protection Agency, in cooperation with the Canada Centre for Remote Sensing, is developing and distributing a time series of Landsat Multi-Spectral Scanner data (see Chapter 9) for all of North America, suitable for use in analyses of land use change (US EPA 1993). The images were acquired in 1973, 1986, and 1991 (±one year), and are georegistered and radiometrically and geometrically corrected (see Chapter 9). The images will be available at the cost of duplication from the EROS Data Center, and metadata describing the scenes can be queried via the USGS Global Land Information System (GLIS).

Soils. The FAO/UNESCO Soil Map of the World (1 : 5 000 000) consists of 18 partly overlapping sheets with accompanying explanatory text and details of the legend criteria (Sombroek & Colenbrander 1990). A digital version is available through UNEP.

2.6.2 Major U.S. databases

A number of U.S. government agencies have developed or are developing digital databases for GIS use (Table 2.8). Federal agencies in the process of digitizing maps include the USGS (topographic maps, land use maps), the U.S. Natural Resources Conservation Service (soil surveys), the U.S. Fish and Wildlife Service (wetland inventories), and the U.S. Census Bureau (census data). As GIS use becomes more widespread, these digital databases will become increasingly available.

A base map containing georeferencing information and standard features such as roads and streams is fundamental to most GIS applications. This information can be obtained in vector format from the USGS's Digital Line Graph (DLG) series. DLG databases are derived from USGS

Table 2.8 Some sources of digital GIS databases in the U.S.

Database	Source	Scale	Unit area covered
Digital line graph (DLG)	USGS	1 : 24 000	7.5 × 7.5′
Digital line graph (DLG)	USGS	1 : 100 000	7.5 × 7.5′ (hydrography) 30 × 30′ (other)
Digital line graph (DLG)	USGS	1 : 2 000 000	21 regions cover the U.S.
Digital elevation model (DEM)	USGS	1 : 24 000	7.5 × 7.5′
Digital elevation model (DEM)	Defense Mapping Agency	1 : 250 000	1 × 1°
Digital raster graphics (DRG)	USGS	1 : 24 000	7.5 × 7.5′
Digital orthophoto quads (DOQ)	USGS	1 : 24 000	7.5 × 7.5′
Land use/land cover (LULC)	USGS	1 : 100 000 1 : 250 000	30 × 30′ 1 × 2°
Geographic names information system (GNIS)	USGS	N/A	7.5 × 7.5′
Soil survey (SSURGO)	Natural Resources Conservation Service	1 : 15 840– 1 : 31 680	U.S. Counties
State soils (STATSGO)	Natural Resources Conservation Service	1 : 250 000	U.S. States
TIGER files	U.S. Census Bureau	1 : 100 000	U.S.

maps, and consist of several different thematic layers. DLG files are available from EROS Data Center at scales of 1 : 2 000 000 for state and national applications (Fig. 2.13), 1 : 100 000 for state and regional applications, and 1 : 24 000 for site-specific applications. The 1 : 2 000 000 and 1 : 100 000 series are available for the entire U.S., but the 1 : 24 000 series is available currently for only part of the country. A useful companion to DLGs is the Geographic Names Information System (GNIS), which is the official repository for U.S. physical and culture place-names.

Another digital product useful in ecological research is the DEM. DEM databases produced by the USGS are raster representations of elevation, with each pixel assigned an elevation value. USGS 1 : 100 000 DEMs are available for the entire country, and 7.5-minute DEMs are

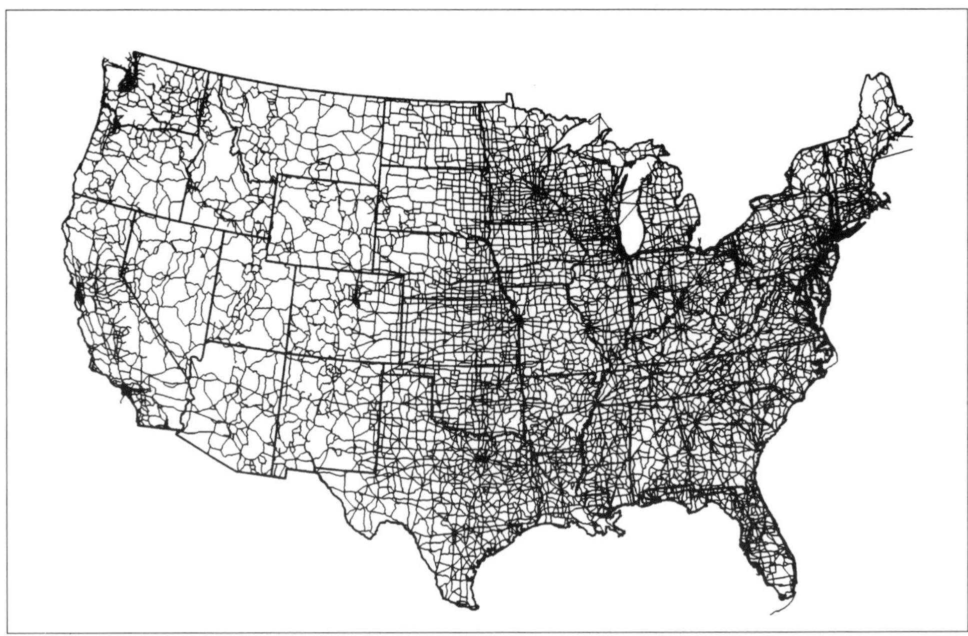

(a)

(b)

Fig. 2.13 GIS layers compiled for the conterminous U.S. from 1 : 2 000 000 USGS DLG databases: (a) roads and trails, (b) perennial streams. Courtesy of James Westman, AScl Corporation, U.S. EPA Environmental Research Laboratory, Duluth, Minnesota.

available for portions thereof. A current status map for 7.5-minute DEMs can be viewed from the USGS WWW Home Page (see Appendix).

Digital Raster Graphics (DRG) and Digital Orthophoto Quadrangles (DOQ) are USGS products that provide important background information for ecological studies. A DRG is an image of a USGS standard series topographic map, scanned at a minimum resolution of 250 dots per inch, and georeferenced to the UTM projection. A DOQ is a digital image of aerial photographs in which displacements caused by the camera and the terrain have been removed, combining the image characteristics of a photograph with the geometric qualities of a map.

The USGS Land Use and Land Cover (LULC) 1 : 250 000-scale maps were derived from aerial photograph interpretation, and classify the U.S. into nine categories: built-up land, agricultural land, rangeland, forest land, water, wetlands, barren land, tundra, and perennial snow or ice. LULC maps are available in digital form, but are coarse and outdated in areas of rapidly changing land use.

Several U.S. agencies are collaborating to develop a contemporary nationwide land cover data layer for GIS, called the Multi-Resolution Land Characteristics (MRLC) Monitoring System. Land cover classification will be done from Landsat Thematic Mapper and AVHRR imagery (see Chapter 9), in combination with various ancillary data sets (Steyaert 1996).

2.6.3 GIS data on the Internet

The Internet is the world's largest computer network, providing rapid worldwide access to information about everything, including GIS. GIS software, databases, satellite imagery, articles, and information can be downloaded via the Internet. GIS users can communicate with each other over the Internet, and ask questions or give opinions about GIS topics. Academic users can usually access the Internet via an Ethernet or other network link with their University computing center, and commercial and private users can access it through commercial communications services.

"Surfing" the Internet through the use of a search program is a good way to explore its capabilities. There are many on-line resources and **FAQs** (Frequently Asked Questions) online that can answer basic questions about the Internet, a few of which are listed in the Appendix. The advantage of using the Internet to find out about the Internet is that it is always up to date.

2.6.3.1 Information presentation systems

The World Wide Web (WWW) is rapidly gaining popularity as an Internet information presentation system, replacing older systems such as Gopher and WAIS (Wide Area Information Servers). An information presentation system is based on client–server interactions, where software on your computer (the client) makes a request to a server on another computer which answers your request (Krol 1992). The systems differ in format (e.g. menus vs. hypertext windows) and types of data accessed:

WWW — a system for viewing Internet pages that are written in hypertext markup language (HTML). WWW pages typically contain color graphics and hyperlinked text, a method of presenting information where highlighted words in the text can be expanded to provide additional information.

Gopher — a series of subject menus for browsing Internet resources pertaining to a menu topic, and retrieving applicable files and information. Data accessed by gopher include menus, documents, indices, and Telnet connections.

WAIS — a service that allows you to search indexed documents on the Internet using key words or phrases. Data searched by WAIS are always indices, and data returned from the index are always documents.

2.6.3.2 Standard Internet access tools and terms

E-mail a way to send and receive messages to other networked computer users.

Telnet a program that lets you connect to another computer; also the command used to initiate that program.

ftp (file transfer protocol) a program that lets you copy a file from another computer onto your own computer.

anonymous ftp allows you to log in anonymously onto another computer on the Internet, and retrieve publicly accessible files.

HTML (Hypertext Markup Language) a computer language used to code documents for use on the WWW.

VRML (Virtual Reality Markup Language) a computer language which concisely describes complex virtual objects (e.g. spheres, cones, cubes) or worlds (e.g. digital terrain). VRML conforms to HTML standards, and is being increasingly used on the WWW.

http (hypertext transfer protocol) the protocol for transfering HTML files.

hyperlink underlined or highlighted words or icons that, when clicked on with a computer mouse button, provide additional information by connecting with other WWW pages.

browser an interface program to display WWW pages and follow hyperlinks. The two most widely used WWW browsers are **Mosaic**, developed by the U.S. National Center for Supercomputing Applications (NCSA), and **Netscape Navigator**. Versions of Netscape Navigator are available for the PC, for the Macintosh, and for various Unix computers.

URL (Uniform Resource Locator) an Internet address. URLs for WWW sites begin with http:// .

CHAPTER 3

Basic GIS operations

The GIS analytical capabilities described in this chapter are divided into three general groups: database operations that may be performed on spatial or non-spatial data, operations performed on individual spatial data layers, and operations performed on multiple spatial data layers (Fig. 3.1). Despite the differences in data structure between raster and vector systems, most of these operations can be done with either data structure.

The operations in raster GIS are often grouped differently, into **local**, **focal**, and **zonal** operations (Tomlin 1990). Local functions are those that compute a new value for each location as a function of existing data explicitly associated with that location in single or multiple layers (Fig. 3.2). Focal functions are those that compute a new value for each location in a layer as a function of the spatial relationship between the focal cell and its **neighborhood**, a neighborhood being a two-dimensional area that bears a specified distance and/or directional relationship to the focal cell within. The most frequently used neighborhood is a 3×3 cell square surrounding the focal cell (Fig. 3.3). Zonal functions are those that compute a new value for each location as a function of existing values associated with a zone containing that function (Fig. 3.4). **Zones** represent two-dimensional areas but unlike neighborhoods, zones may vary in size and shape, and may not overlap. Whereas each location in a layer can be part of only one zone, it can be part of any number of neighborhoods (Tomlin 1990). Although this system of organization is not used here, these terms will be referred to in descriptions of several raster GIS functions.

In addition to the basic GIS functions described in this chapter, more specialized GIS functions are described in subsequent chapters.

3.1 Database operations performed on spatial or non-spatial data

3.1.1 Data selection and query

Like other database management systems, a GIS can be used to select data that meet certain criteria. Although simple in concept, this means of filtering out extraneous information is the GIS analysis tool used most often. Selection of database records may be done: (i) interactively; (ii) with a look-up table; (iii) by specifying numerical thresholds; or (iv)

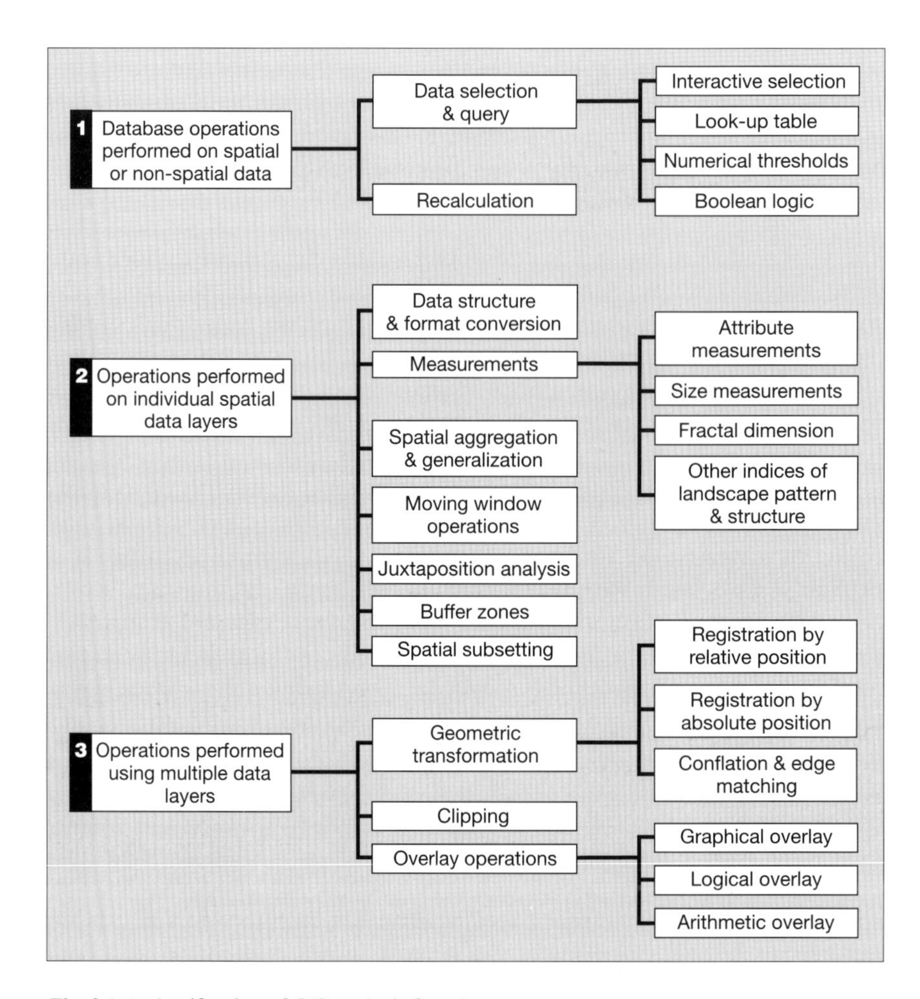

Fig. 3.1 A classification of GIS analysis functions.

through Boolean logic. Simple examples of these selection processes are provided below.

Modern GIS are coupled with database management systems (DBMS) which provide a wider range of selection and query tools than can be summarized here. Special query languages are often used, such as Structured Query Language (SQL). Linkages with graphical statistical packages, such as S+, have also facilitated analysis of the data generated by GIS.

3.1.1.1 Interactive selection

Interactive selection is used for simple database query, such as selecting records for display to the computer monitor. Let us consider a data layer that contains information about macroinvertebrates inhabiting the Pine and Chippewa Rivers of Michigan (Table 3.1). Macroinvertebrate sam-

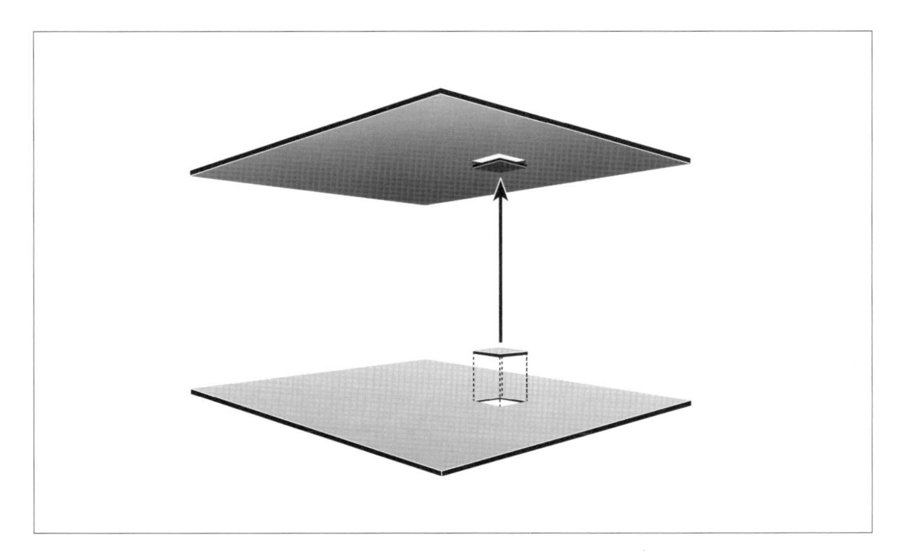

Fig. 3.2 Local functions. From *Geographic Info Systems and Cartographic Modeling* by Tomlin, © 1990. Reprinted by permission of Prentice-Hall, Inc., Upper Saddle River, NJ.

ples were obtained by placing artificial substrates (masonite Hester–Dendy samplers) at numerous points in the rivers, and allowing macroinvertebrate colonization for 7–8 weeks prior to sampler retrieval (Arthur *et al.* 1996). A unique code identifies each sample in the database (Table 3.1), as well as its corresponding location in the GIS data layer

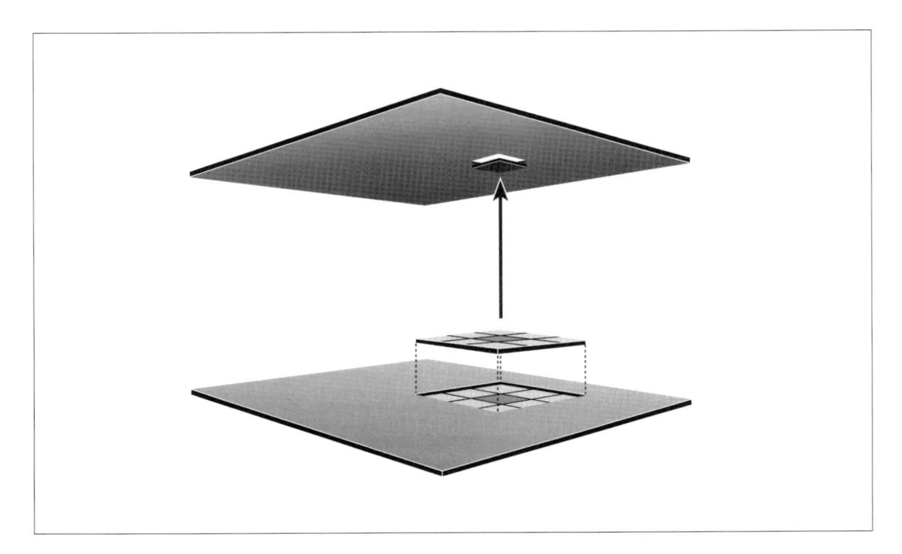

Fig. 3.3 Focal functions. From *Geographic Info Systems and Cartographic Modeling* by Tomlin, © 1990. Reprinted by permission of Prentice-Hall, Inc., Upper Saddle River, NJ.

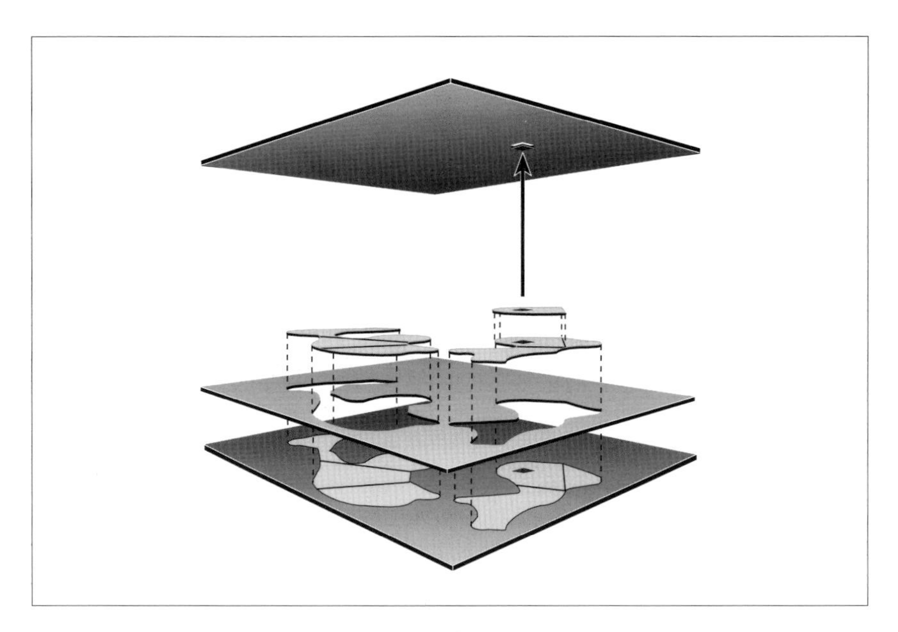

Fig. 3.4 Zonal functions. From *Geographic Info Systems and Cartographic Modeling* by Tomlin, © 1990. Reprinted by permission of Prentice-Hall, Inc., Upper Saddle River, NJ.

Table 3.1 Macroinvertebrates colonizing artificial substrate samplers placed in the Chippewa and Pine Rivers.

Sample ID	No. in sample	Richness*	No. of EPT taxa†	Majority genus	Family of majority genus	Functional group of majority genus
250–5	1725	42	12	*Tanytarsini*	Chironomidae	Collector
255	1206	49	17	*Stelechomyia*	Chironomidae	Collector
250–3	1888	44	16	*Stelechomyia*	Chironomidae	Collector
256	899	40	15	*Hydropsyche*	Trichoptera	Collector
253	540	24	10	*Tanytarsini*	Chironomidae	Collector
254	1189	36	14	*Stenonema*	Ephemeroptera	Scraper
252	619	41	11	*Polypedilum*	Chironomidae	Predator
251	697	37	16	*Stenonema*	Ephemeroptera	Scraper
250	800	37	14	*Stenonema*	Ephemeroptera	Scraper
240–3	1865	44	12	*Cladopelma*	Chironomidae	Collector
242	964	36	15	*Stenonema*	Ephemeroptera	Scraper
240–2	1707	46	15	*Cladopelma*	Chironomidae	Collector
241	2635	40	10	*Neureclipsis*	Trichoptera	Collector
240–0	2048	38	10	*Tricorythodes*	Ephemeroptera	Collector
240	1233	31	13	*Stenonema*	Ephemeroptera	Scraper

Data summarized from Arthur *et al.* 1996. * Richness = total number of taxa in sample.
†EPT taxa = Ephemeroptera–Plecoptera–Trichoptera taxa.

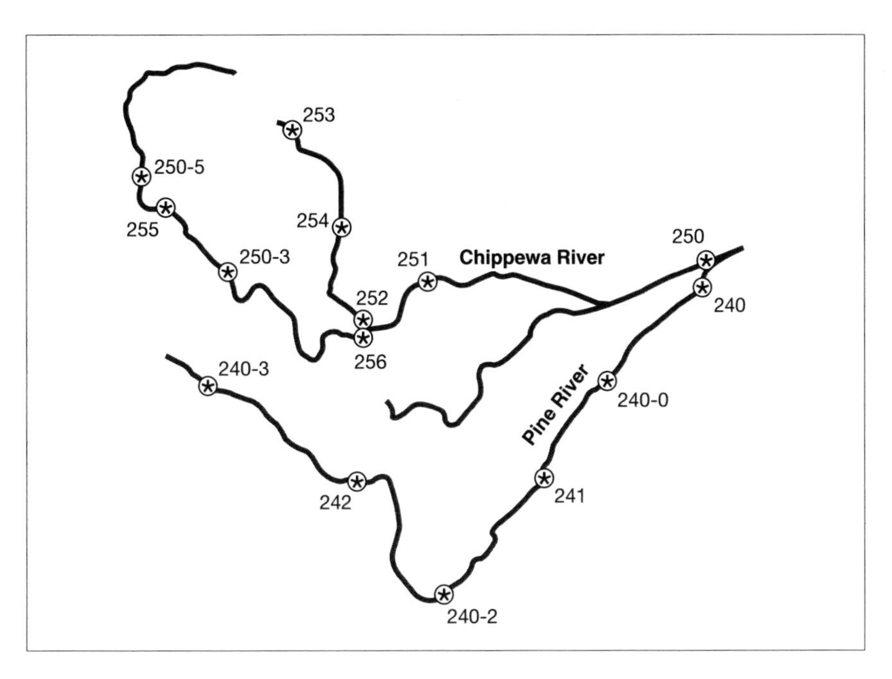

Fig. 3.5 Macroinvertebrate sampling stations on the Pine and Chippewa Rivers, Michigan. After Arthur *et al.* 1996.

(Fig. 3.5). Assuming that each sample is representative of conditions upstream as far as the next sample point, we can use these data to classify river segments by the majority genus of their macroinvertebrate inhabitants (Fig. 3.6). This results in a complex map, requiring nine different symbols for each of the river classes: eight genera classes and one "unknown" class. The unknown class is used for a tributary that was not sampled and for portions of the rivers downstream from the lowest sample points.

We could simplify this map by interactively selecting only those river segments having *Stenonema* as the majority genus, and assigning them a new symbol for redisplay. In this simple **reclassification**, the number of input and output categories is small, and thus conducive to interactive selection.

3.1.1.2 Look-up table

A look-up table that equates initial data values with new data values is more suitable for complex selections, or for simple selections that are done repeatedly. A look-up table does not involve mathematical operations, but merely renumbers each data record, often for the purpose of **generalization**. For example, the information in Table 3.1 could be used

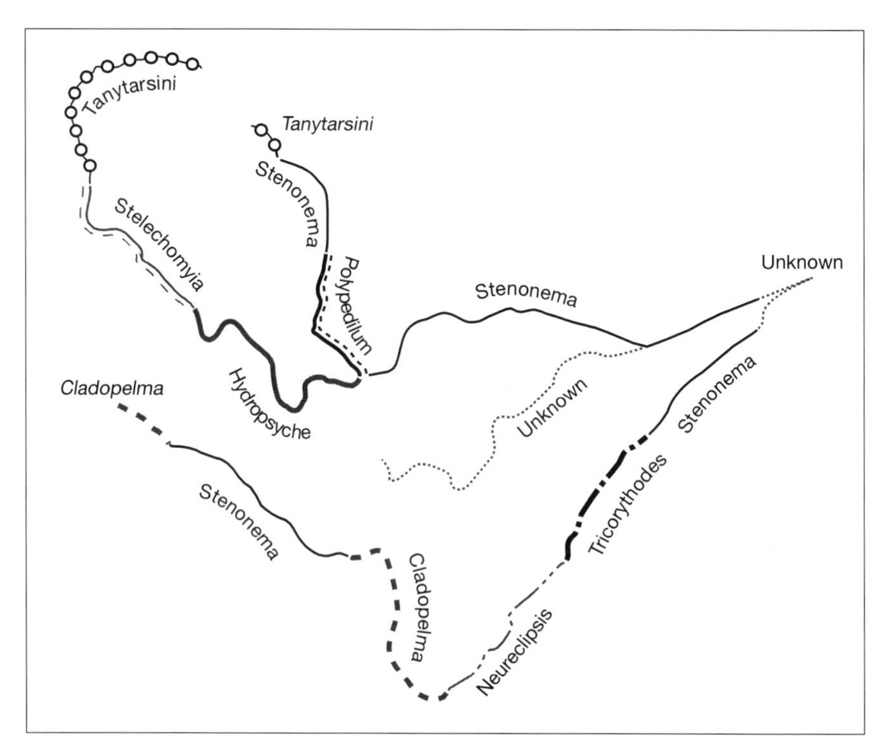

Fig. 3.6 River segments classified by majority macroinvertebrate genus.

to reclassify the river segments according to the family of the majority genus, reducing the number of classes to four: Chironomidae, Ephemeroptera, Trichoptera, and unknown. The computer would use the information in this look-up table to reclassify the initial data categories to the new, simplified categories (Fig. 3.7). Alternatively, the computer could use the look-up table to reclassify river segments by the feeding function of the majority genus (Merritt & Cummins 1984): collector, predator, scraper, or unknown (Table 3.1).

Although data can be generated by reclassification, they cannot be made more detailed. A database reclassified from a source database containing eight classes will have eight or fewer classes; a ninth cannot be added (except for an "other" or "unknown" category) without obtaining additional data. Therefore, it is important to begin with a detailed classification system, and use reclassification to generalize as needed.

3.1.1.3 Numerical thresholds

Quantitative data can be selected by a numerical threshold. For example, a threshold value for the number of Ephemeroptera, Plecoptera, and Trichoptera (EPT) taxa in the macroinvertebrate sample (Table 3.1) could

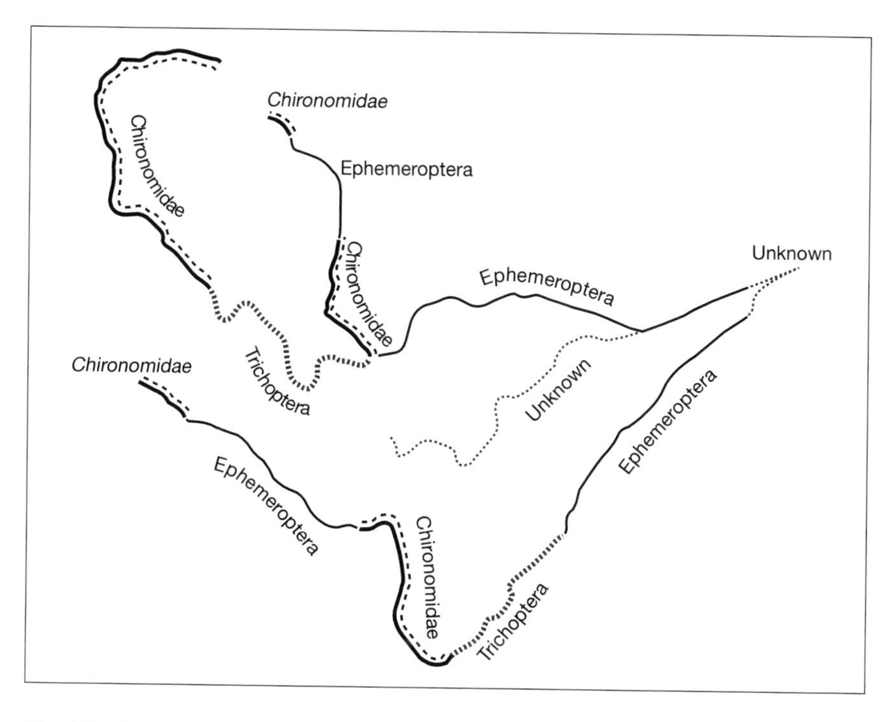

Fig. 3.7 River segments classified by the family of the majority macroinvertebrate genus.

be used as an index of community quality to distinguish river segments with many EPT taxa (high quality) from those with few (low quality).

Quantitative data can also be divided by multiple thresholds, such as dividing the range of numerical values into serial classes with equal arithmetic intervals (see Chapter 2). Figure 3.8 illustrates such a classification using species richness data from Table 3.1. Species richness is a count of the total number of taxa in a sample, a measure of biodiversity, so one or more of these data ranges could be selected to highlight river segments with the greatest biodiversity.

3.1.1.4 Boolean logic

Boolean logic uses the following operators to perform operations on two or more data sets (Fig. 3.9):

AND — intersection
OR — union
NOT — negation
XOR — exclusionary or.

Boolean operators select data records based on two or more attributes. For example, if A is "majority genus = *Stenonema*," and B is "no. of EPT

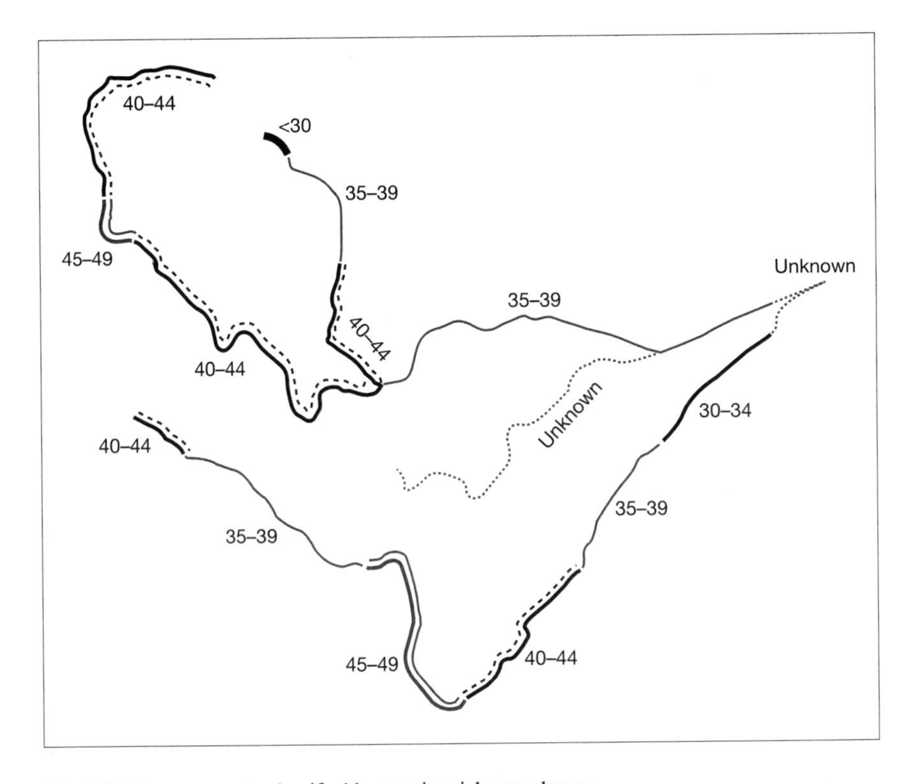

Fig. 3.8 River segments classified by species richness classes.

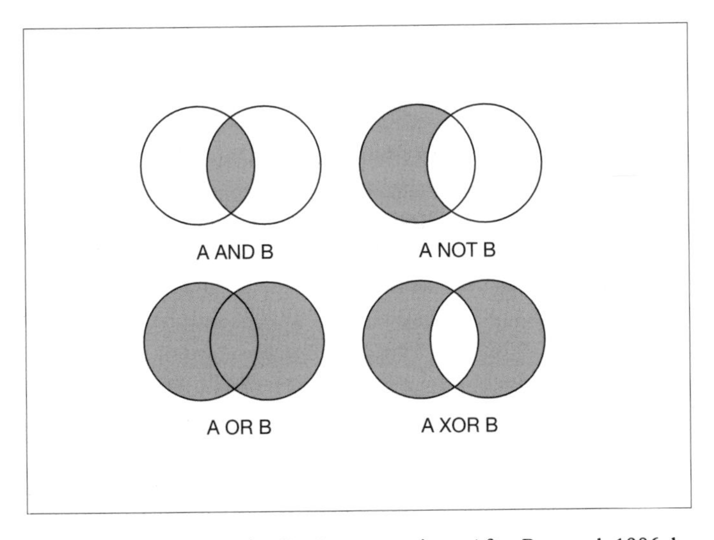

Fig. 3.9 Venn diagrams representing Boolean operations. After Burrough 1986, by permission of Oxford University Press.

Table 3.2 Boolean operations applied to the data in Table 3.1.

Boolean operation	Sample ID of river segments for which the statement is true
A AND *B*	242, 251
A NOT *B*	240, 250, 254
A OR *B*	240, 240-2, 242, 250, 250-3, 251, 254, 255, 256
A XOR *B*	240, 240-2, 250, 250-3, 254, 255, 256

taxa ⩾ 15," then results of the different Boolean operations would be as follows:

A AND *B* = river segments dominated by *Stenonema* that have ⩾ 15 EPT taxa

A NOT *B* = river segments dominated by *Stenonema* that have < 15 EPT taxa

A OR *B* = river segments dominated by *Stenonema* or having ⩾ 15 EPT taxa

A XOR *B* = river segments either dominated by *Stenonema* or having ⩾ 15 EPT taxa, but not both

Table 3.2 lists the individual river segments for which these conditions would apply.

The examples in Table 3.2 illustrate the use of Boolean logic with classical set theory, in which a feature either is or is not a member of a set. However, Boolean operations can also be applied to fuzzy sets, in which elements of the universe have grades of membership in a set (see Chapter 7).

3.1.2 Recalculation

The previous examples used simple data retrieval tools to select records having desired attributes. **Recalculation** involves the use of mathematical expressions to derive new attribute data from one or more existing attributes. For example, if experimental data revealed that the artificial substrate samplers underrepresented species richness by 12%, we could multiply the species richness data in Table 3.1 by 1.12 to correct for this sampler-induced error. In a raster GIS, recalculation is performed as a "local" function on the data in each cell of an individual data layer.

Recalculation is often used in a GIS to transform data to unit length or area. For example. to calculate the gradient of the river segments shown in Fig. 3.5, we would divide the difference in elevation between sample stations by the length of river separating them. Other recalculation operations include: add constant, subtract constant, multiply by constant, divide by constant, exponentiate, trigonometric functions, and logarithmic functions.

3.2 Operations performed on individual spatial data layers

3.2.1 Data structure and format conversion

GIS data come in a variety of formats. The differences between raster and vector data structures have already been discussed (Chapter 2), but different formats exist even within these broad categories. Some of these differences are imposed by the agencies responsible for generating generic data sources, whereas others are imposed by the format output capabilities of different GIS software packages. Berger and co-workers (1996) provide an example in which generic digital elevation data had to be modified using two different software programs before the data could be read into a visualization program. Whatever the reason, transformation of data format can be difficult and time-consuming, and the initial appeal of using existing data layers is often tarnished by the reality of this difficulty.

The ability to read and transform a variety of data formats is an essential characteristic of a good GIS, an ability that is not evaluated easily from vendor claims. Personnel experienced in the intricacies of different data formats are invaluable in this process. Posting questions on Internet bulletin boards or list services that are subscribed to by GIS practitioners is also a good way to find answers to data formatting problems.

3.2.2 Measurements

A variety of measurements can be performed on spatial data in a GIS, ranging from simple area measurements to more complex landscape indices. Like the database operations already discussed, none of these measurements alters or generates spatial features.

3.2.2.1 Attribute measures

Mathematical summaries of attribute values are often more useful than the individual data records, particularly when a database is very large, as is often the case with GIS databases. Unlike recalculation, in which a new attribute is generated by performing the same mathematical operation for each record, descriptive statistics mathematically summarize the attribute data contained in a range of records. Categorical and ordinal data are summarized by determining the frequency of (i.e. counting) the occurrences of different attribute classes (F_i, $i = 1 \ldots t$), where t is the total number of attribute classes. Frequency statistics can be computed for continuous data by grouping data classes into serial intervals ("bins") and counting the number of data records that fall within each interval. Frequency statistics often are displayed graphically as histograms (see Fig. 2.7).

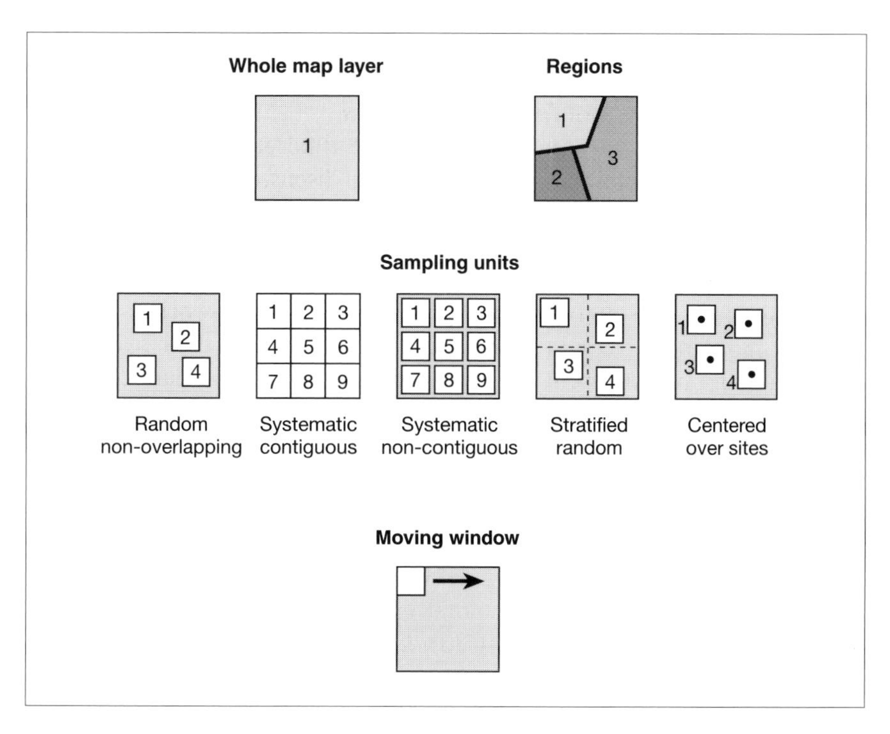

Fig. 3.10 Spatial configurations for placement of sample areas for calculating summary statistics using the r.le programs. From Baker and Cai 1992, by permission of SPB Academic Publishing.

Descriptive statistics can be calculated for an entire map or a user-defined spatial subset circumscribing a smaller region of the map (e.g. a biological preserve). In a raster GIS, descriptive statistics may also be computed for rectangular samples of the grid cells in the map. A variety of spatial configurations may be used for placement of sample areas (Fig. 3.10). The advantage of this approach is reduced computing time and the ability to use univariate statistical analyses to compare groups of sampled data from different maps or regions within a map.

The r.le programs written for the GRASS raster GIS (Baker *et al.* 1993) compute attribute measures for interval or ratio data (see Chapter 2) within entire maps, regions of maps, or map samples:

1 Mean pixel attribute ($\bar{x}$) = the average value of the attributes of all the pixels in the sampling area. Each attribute is weighted by how many pixels it occupies:

$$\bar{x} = \frac{\Sigma_{i=1}^{t} (F_i * i)}{N} \tag{3.1}$$

where N is the total number of pixels sampled. This type of measure might be used to compare average greenness indices (NDVI) for a time series of satellite images (see Chapter 9).

2 Standard deviation of pixel attributes.

3 Mean patch attribute, calculated by summing the attribute values of each patch and dividing by the number of patches. This type of measure might be used to compare average elevation of wetland patches within different geomorphic regions on a map.

4 Standard deviation of patch attributes (s), calculated as:

$$s = \sqrt{\frac{\Sigma_{i=1}^{m} (x_i - \bar{x})^2}{m}} \tag{3.2}$$

where x_i is the attribute of patch i, and m is the number of patches.

The capability to reduce the data in a map to a few summary statistics benefits ecological modelers, who use the numerical output from GIS to derive mathematical relationships between ecological variables. The data from a statistical summary can be exported to statistical and modeling programs for further analysis.

3.2.2.2 Size measurements

Area measurements are standard and straightforward in vector and raster GIS. Areas are computed in a raster database by counting the number of cells with a given value and multiplying by the cell area. Area overestimates may occur in a raster GIS when the grid cell size is much larger than the area of objects represented, such as a 30 × 30 m cell representing a 1-m^2 water well. The most common area measurements generated by a GIS are the total area of different attribute classes, but measures of the average and standard deviation of patch area by class are also used in landscape ecology.

Distance measures are used to determine the distance between two points, or the length of a straight line segment. In landscape ecology, inter-patch distances are measured between the **centroids** of polygons, computed in a raster GIS as the average row and column of the cells that are part of the patch. Alternatively, inter-patch distances can be measured from nearest edge to nearest edge or from centroid to nearest edge. Distance measures are also used to determine the longest axis across a patch.

Distance measures are computed in a vector GIS using the Cartesian formula for the distance between two points:

$$d = \sqrt{(x_1 - x_2)^2 + (y_1 - y_2)^2} \tag{3.3}$$

where d is distance, x_1 and y_1 are the x,y coordinates of the first point, and x_2 and y_2 are the x,y coordinates of the second point. Individual segment lengths are summed to derive the cumulative length of strings (see Chapter 2). String length and polygon perimeter data commonly are provided by attribute tables in vector GIS.

Length measurements are more complicated in a raster GIS. Straight lines which parallel a row or column in the raster array have a length equal to the number of cells multiplied by the pixel dimension. For example, a straight line of 43 cells, each representing a 7×7 m square, would have a length of 301 m (43 cells $\times$ 7 m/cell). A straight line of 43 cells along the diagonal of the same data layer would have a length equal to 425.7 m (i.e. 43 cells $\times$ 9.9 m, the length of the cell diagonal). Other linear measures would fall between these two extremes, depending upon the orientation and complexity of the line. For example, each pixel in a rasterized hydrologic database represented 7.61 m of stream channel, 9% longer than the 7 m pixel dimension which would have been obtained for a straight stream paralleling the grid structure (Johnston & Naiman 1990). Estimated stream lengths obtained using this coefficient were within 3.2% of stream lengths measured using Cartesian coordinates. Such empirically derived coefficients, however, are applicable only to linear features of similar complexity.

Perimeter measures are also more complex in a raster GIS than they are in a vector GIS, because patch edges must be distinguished from patch interiors before they can be measured (see Section 3.2.5).

3.2.2.3 Fractal dimension

In his paper, "How long is the coast of Britain?," Benoit Mandelbrot (1967) showed that for most naturally occurring phenomena, the amount of resolvable detail is a function of scale, and that increasing the map scale reveals additional complexity. Mandelbrot (1983) developed this idea into the theory of fractals, which has been applied in fields as diverse as ecology, geography, and geophysics (Burrough 1984; Milne 1991; Lam & DeCola 1993).

Dimensions are essential for representing geometric objects (Milne 1991). For example, when the area of a disk $= \pi r^2 = \pi r^d$ is calculated, the radius is squared because the disk is a two-dimensional set of points (i.e. $d = 2$). Points, lines, planes, and solids have topological dimensions that are integers (i.e. $d = 0$, 1, 2, and 3), whereas **fractal dimensions** may be non-integers.

In general, fractal analyses quantify scaling relations, of the form

$$y = x^{\beta} \tag{3.4}$$

of a landscape property y over a range of scale x (Ritters *et al.* 1995). In practice, β is estimated by the slope of a double logarithmic linear regression of the form $\ln(y) = \alpha + \beta\ln(x)$, and the fractal dimension is then found by transforming β (Falconer 1990).

The fractal dimension of linear strings can be described as the relationship between a quantity Q and the reference length scale L over which Q is measured (Stanley 1986).

Thus:

$$Q(L) = L^{D_q} \quad (3.5)$$

where D_q represents a fractal dimension for the quantity Q. To picture this, imagine a trail of mouse tracks in the snow between a woodpile and a food cache. If the mouse walks in a very straight line, then the distance traveled is equal to the reference distance L between the woodpile and the food cache, and $D_q = 1$. If the mouse's path is random, then $D_q = 2$ (Stanley 1986). Weins and Milne (1989) found that beetle (*Eleodes longicollis*) paths were fairly direct ($D_q = 1.1$), indicating that beetles do not behave randomly in their movement. Figure 3.11 illustrates linear functions having fractal dimensions varying from $D = 1.1$ to $D = 1.9$ (Burrough 1986).

In addition to quantifying the complexity of linear features, fractal dimension can also be used to characterize the complexity of two-dimensional patches. In Euclidean geometry the area A of a patch is related to the diameter L by the relationship

$$A = BL^2 \quad (3.6)$$

For disks, B is a constant equal to $\pi/4$. The relationship is generalized to fractal patches by the relationship between area and patch length:

$$A = \beta L^{D_a} \quad (3.7)$$

where the reference length scale (L) is the distance between the two most distant points on the perimeter, and D_a is the fractal dimension of area. High dimensions ($D_a \approx 2$) indicate an obtuse shape (Milne 1991).

3.2.2.4 Other indices of landscape pattern and structure

Quantification of spatial pattern and structure is central to the study of landscape ecology, and GIS has contributed greatly to the advancement of the discipline due to the landscape measurement and characterization tools it provides to ecologists (Johnston 1990). In addition to the attribute, size, and fractal measures already discussed, landscape metrics include diversity, texture, juxtaposition, and shape measures. Computer programs such as Fragstats (McGarigal & Marks 1994) and r.le (Baker & Cai 1992) have simplified calculation of such metrics.

The ease of measurement provided by GIS has resulted in a proliferation of landscape metrics, many of which are duplicative. Riitters *et al.* (1995) evaluated metrics of landscape pattern and structure to identify those that were not statistically redundant. Starting from 55 candidate metrics calculated for 85 raster land use and land cover maps, they

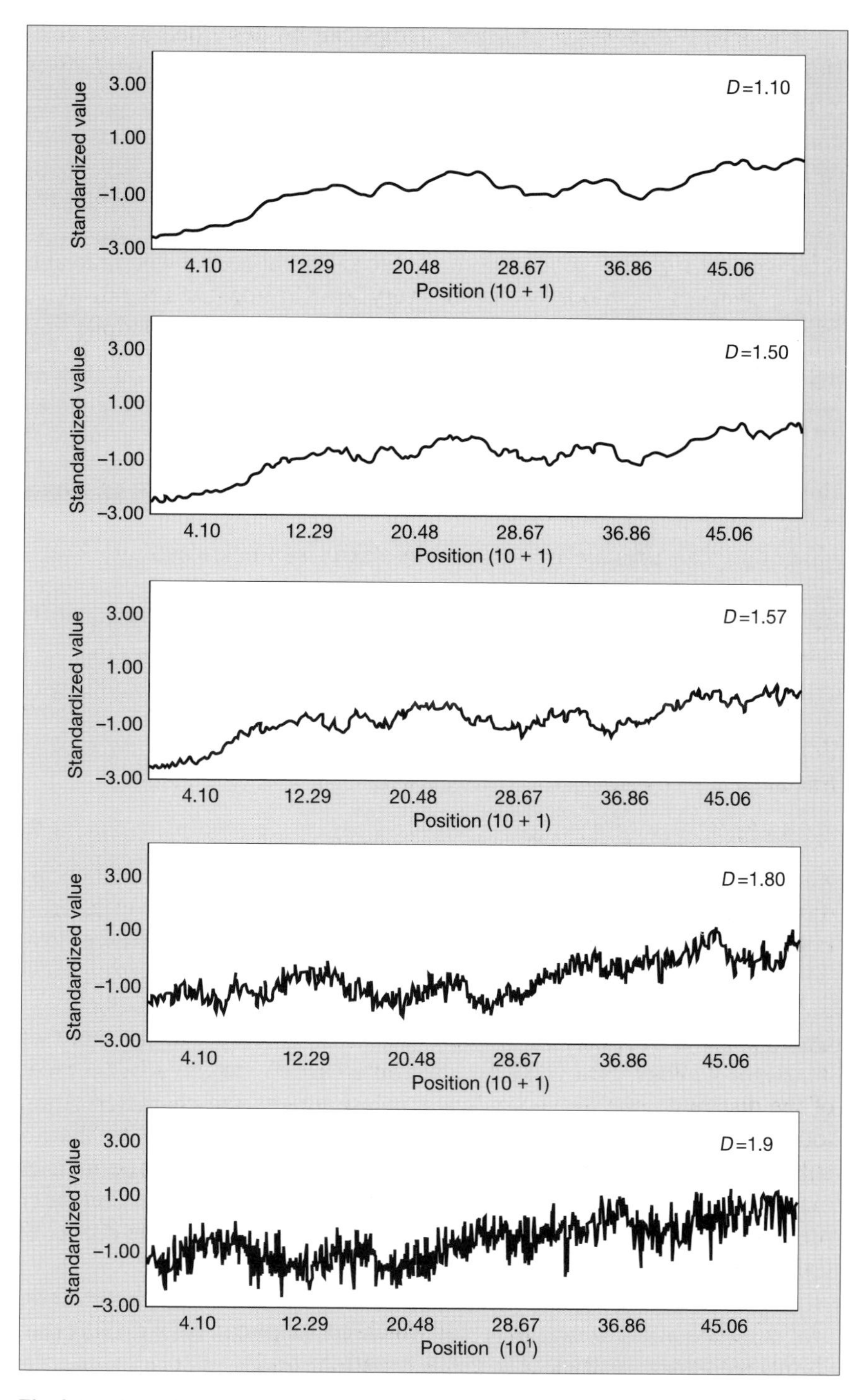

Fig. 3.11 Linear functions having different *D* values. From Burrough 1986, by permission of Oxford University Press.

identified 26 metric groups for which all within-group correlations were larger than +0.9 (Table 3.3), and then used multivariate factor analysis to identify a smaller number of apparently independent axes. Univariate metrics having the highest correlations with these axes were selected to represent them, five of which are presented below.

The first axis was termed **average patch compaction**, and was correlated with average perimeter-area ratio (PA):

$$PA = \frac{1}{m} \Sigma_{k=1}^{m} \frac{OE_k}{S_k} \tag{3.8}$$

where: m is the total number of patches; S_k ($k = 1 \ldots m$) is the number of cells of the k-th patch; and OE_k is the number of edges enclosing a patch.

Table 3.3 Groups of metrics used in factor analysis by Riitters *et al.* (1995), reprinted by permission of SPB Academic Publications.

Group representative metric	Reference
1 Number of attribute classes	
2 Shannon evenness of attribute classes	Equation 3.10
3 Kempton–Taylor Q-statistic	Magurran 1988
4 Shannon contagion	Equation 3.9
5 Sum of diagonal elements of adjacency matrix	
6 Average attribute class lacunarity from the scaling of attribute density with neighborhood size	
7 Average proportion of area in patches larger than 5 cells	
8 Perimeter-area scaling, patch perimeter complexity	
9 Perimeter-area scaling, patch topology transformation, enclosing cells basis	Equation 3.13
10 Patch area-bounding circle scaling	
11 Patch perimeter complexity from the scaling of Euclidean distance to actual distance along large patch perimeters	Weins & Milne 1989
12 Metric of large-patch "mass" from the scaling of patch density with neighborhood size	Voss 1988
13 Average large-patch lacunarity from the scaling of patch density with neighborhood size	
14 Number of patches	
15 Average patch size or area	
16 Average patch radius of gyration	Pickover 1990
17 Average number of inside edges per patch	
18 Average patch perimeter-area ratio	Equation 3.8
19 Average patch adjusted perimeter-area ratio	Baker & Cai 1992
20 Average patch normalized area, square model	Equation 3.12
21 Average patch topology ratio	
22 Average patch ratio of number of inside edges to area	
23 Average patch adjusted perimeter-area ratio	
24 Average ratio of patch area to area of the circumscribing circle	Baker & Cai 1992
25 Average ratio of patch radius of gyration to long axis length	
26 Average patch ratio of perimeter cells to perimeter edges	

Landscape patches tend to be compact on maps with low *PA* values and convoluted on maps with high *PA* values. Minimum and maximum *PA* values from 85 maps studied by Riitters *et al.* (1995) were 1.051 and 1.785, respectively.

The second axis was related to **map texture**. The univariate metric representing this axis is the Shannon contagion index (O'Neill *et al.* 1988; Li & Reynolds 1993):

$$SHCO = 1 - \frac{SHHO}{2\ln(t)} \tag{3.9}$$

where t is the number of attribute classes and *SHHO* is the Shannon homogeneity of the adjacency matrix (Shannon & Weaver 1949), calculated as:

$$SHHO = -\Sigma_{i=1}^{t}\,\Sigma_{j=1}^{t}\,[v_{ij}\ln(v_{ii})] \tag{3.10}$$

where v_{ij} is the probability of a grid cell of attribute i being found adjacent to a grid cell of attribute j, calculated as:

$$v_{ij} = \frac{A_j}{\Sigma_{i=1}^{t}\,\Sigma_{j=1}^{t}\,A_{ij}} \tag{3.11}$$

and A_{ij} denotes the frequency of attribute class i being located adjacent to attribute class j in a cardinal direction. Because each edge is counted only once, the adjacency matrix A is without regard to the ordering of the two cells that define an edge. A map with low values of *SHCO* tends to be fine-grained (dissected) and a map with high values of *SHCO* tends to be coarse-grained (clumped). Minimum and maximum observed values in the Riitters *et al.* (1995) study were 0.488 and 0.951, respectively.

The third axis, **average patch shape**, was most correlated with those average patch metrics which employ standardization to an assumed patch shape, such as "average normalized area, square model" (*NASQ*):

$$NASQ = \frac{1}{m}\,\Sigma_{k=1}^{m}\,16\,\frac{S_k}{OE_k^2} \tag{3.12}$$

where OE_k and S_k are as defined in Equation 3.8. *NASQ* has a value of zero for linear patches and one for square patches, and it is sensitive to patch size.

The fourth axis was associated with perimeter-area fractal measures, exemplified by the fractal estimator of patch topology from perimeter-area scaling, enclosing cells basis (*OCFT*):

$$OCFT = \frac{1}{\beta_2} \tag{3.13}$$

where β_2 is the estimated slope from the regression of $\ln(OC_k)$ on $\ln(S_k)$ for all patches with S_k that do not touch the border of the map, where OC_k is the number of cells enclosing a patch. Low values of *OCFT* indicate complicated perimeters, whereas high values indicate simple perimeters; the range of values in the Riitters *et al.* (1995) study was 1.094–1.573.

The fifth axis was most correlated with the **number of attribute classes.** The sixth axis was correlated only with **large-patch density-area scaling**, a fractal metric of large-patch "mass" from the scaling of patch density neighborhood size (Voss 1988). These six axes explained 87% of the variation in the 26 landscape metrics used (Riitters *et al.* 1995).

The landscape indices presented here are not intended to be comprehensive, but are representative of the range of measures available to quantify spatial patterns using a GIS. More detailed discussions of landscape measures are provided by Riitters *et al.* (1995), Magurran (1988), Milne (1991), Musick and Grover (1991), and Gonzales and Woods (1992).

GIS has made it much easier to measure landscape pattern (as indicated by the proliferation of landscape indices), but the significance of those patterns to ecological function (e.g. water quality, wildlife habitat value, ecosystem sustainability) remains an open field of inquiry.

3.2.3 Spatial aggregation and generalization

Spatial aggregation or **convolution** may be used to simplify excess spatial detail in a raster data layer. The user defines the numbers of cells to be aggregated in each $n \times n$ window, or **kernel**. The GIS then divides the entire input data layer into non-overlapping $n \times n$ blocks and analyzes the cell values within each block, deriving a summary value. This summary value is assigned to the new, larger block (Fig. 3.12). Usually, the value assigned to the aggregated block is the majority value (i.e. the value that occurs most frequently within the cells of the block). However, various summary values may be computed for the kernel (Table 3.4).

In addition to reducing file size, this technique may be used to detect ecological patterns at a coarser grain than that of the original file. For example, the 30 × 30 m resolution of Landsat Thematic Mapper Imagery (see Chapter 9) may be suitable for detecting vegetation patterns, whereas a coarser grain created by aggregating TM cells would be more suitable for analyzing bedrock trends.

Data **resampling** is similar, except that the kernel does not have to be a unit number of cells, and may have a different orientation and size than the input cells. Resampling is done to convert a raster data layer to another coordinate systems or projection, or to match raster data layers having different cell sizes prior to overlaying them. Methods for determining which data value should be assigned to the new cell include:

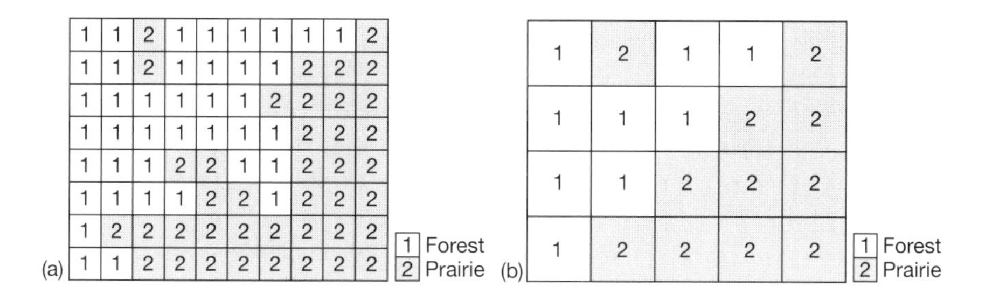

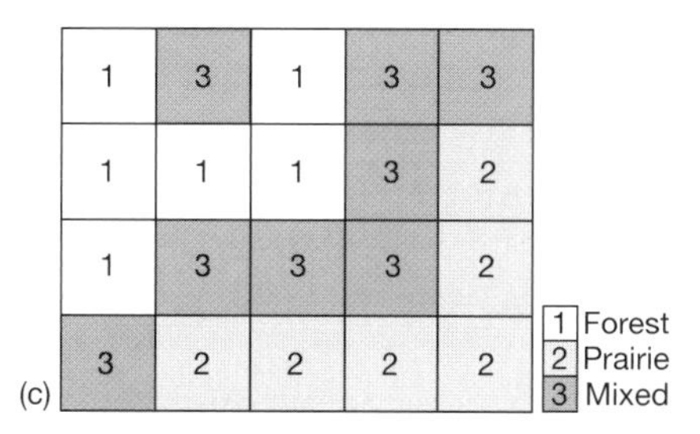

Fig. 3.12 Spatial aggregation of categorical data: (a) original data layer, (b) aggregation based on a majority rule, (c) aggregation with the creation of a new mixed class.

nearest neighbor, **bilinear interpolation**, and **cubic convolution**. The nearest neighbor method assigns the new cell the value of its closest neighbor cell in the original data layer. The bilinear interpolator computes the new cell value from the values of the four cells surrounding the new cell, whereas cubic convolution computes it using a weighted sum of the 16 cells surrounding the new cell.

Clumping is a raster GIS operation that aggregates patches of contiguous cells having the same attribute value. Clumping identifies groups of

Table 3.4 Summary values that can be used in spatial windows.

Summary value	Description
Average	Average of the cell values in the window
Minimum	Minimum of the cell values in the window
Maximum	Maximum of the cell values in the window
Range	Range of the cell values in the window (maximum–minimum)
Median	Median of the cell values in the window
Mode	Mode of the cell values in the window
Majority	Value that occurs in the most cells
Minority	Value that occurs in the least cells
Class count	A count of the number of different classes represented by the cells in the window (a measure of diversity)
Deviation	Standard deviation of the cell values in the window
Proportion	Proportion of the window comprising cell values of a particular class

cells having the same value that touch at a corner or edge, and assigns all of the cells in that group a number that identifies the patch as a unique entity. For example, this process could be used to identify, and uniquely number, individual woodlots in an agricultural landscape. Clumping may be followed by a **sieving** process to eliminate clusters below a minimum size. Once created, the clusters may be classified by size. "Border" or "edge" options in some raster GIS software packages (see Section 3.2.5) create a linear boundary around the clusters identified, which is useful in converting the data from raster to vector format.

Polygon elimination procedures similar to sieving are also done in vector GIS. Polygon elimination can remove very small polygons, as well as **sliver** polygons, long narrow polygons that are often created as artifacts of overlaying two data layers in which polygon boundaries do not match exactly. Sliver polygons are removed by eliminating their longest shared boundary with adjacent polygons.

Area aggregation can also be done in vector GIS by reclassifying polygons so as to generalize their attributes, and dissolving lines that become extraneous as a result of that reclassification. When generalization causes adjacent polygons in a vector database to be assigned the same classification code, such as combining adjacent *Typha* and *Sparganium* stands into a more general "emergent wetland" polygon, the boundaries between them are no longer necessary. These extraneous boundaries do not affect cumulative sums, such as the total area of the new wetland category, but do affect measures of individual patches (e.g. a count of the number of emergent wetland stands). A **dissolve** operation removes these superfluous boundaries, after which topology is rebuilt for the new polygons. When adjacent data layers are **joined** at the edges, a dissolve operation can also be used to remove lines that formerly defined the outer extent of the map.

Lines may also be generalized in a vector GIS by **line coordinate thinning**, which reduces the number of coordinate pairs used to define them. Automated techniques for line coordinate thinning are incorporated into many GIS software packages, based on algorithms developed by Douglas and Peuker (1973).

3.2.4 Moving window operations

The **moving window** is a "focal" function (Fig. 3.3) unique to raster GIS. The data are analyzed for all cells within a two-dimensional window (kernel) containing an odd number of cells (e.g. 3×3, 5×5), and the result is assigned to the central cell at each window position (Fig. 3.13). Each computed value is written to the corresponding cell in the output data layer, and the window is shifted over one position to compute a new value. The process is repeated until the entire data layer has been

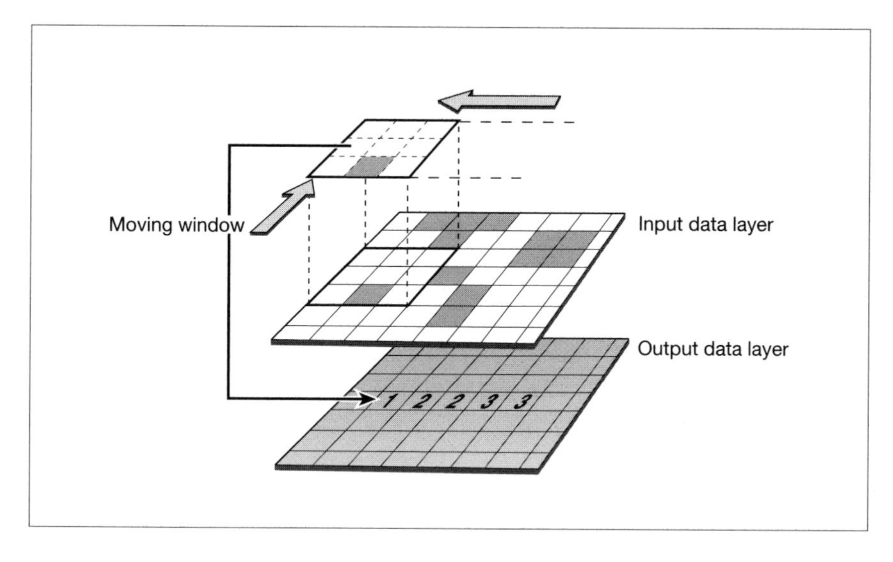

Fig. 3.13 A 3 × 3 moving window. A summary value derived from the values for each of the nine pixels in the scan window is assigned to the central pixel, written to the output data layer, and the window is moved until the entire analysis area has been covered. Adapted from *Geographic Information Systems: A Management Perspective* by Aronoff (1989). WDL Publications. Ottawa, Canada.

scanned and a new data layer created. The new data layer contains the same number of cells as the original data layer, but the data values are the result of the mathematical analysis performed within each kernel. Unlike the systematically contiguous kernels (Fig. 3.10) used in spatial aggregation, moving windows overlap. The analysis neighborhoods of moving windows are usually symmetrical and square, but other shapes can also be used.

The same mathematical functions used in spatial aggregation operations (Table 3.4) can also be used for moving windows. Some functions (e.g. mean, median, majority) can be used to smooth data, reducing the effects of single cells that vary from the general trends. A class count function can be used to determine local diversity. The standard deviation and range functions can be used to enhance the heterogeneity (texture) of an area and to detect edges between adjacent areas of dissimilar class.

Moving window analyses of satellite imagery using a range algorithm (i.e. maximum–minimum value) can be used to detect boundaries by measuring the relative degree of difference between the reflectance values of cells in the window. This technique was used with the normalized difference vegetation index (NDVI), an indicator of net primary productivity derived from satellite image analysis, to detect contrast between areas of high and low productivity (Johnston & Bonde 1989).

Moving windows can be applied to any digital data surface, including aerial photography scanned with a video digitizer or scanning camera. The technique was developed for image analysis and usually applied to continuous data layers, but can also be used with categorical and ordinal raster data layers.

3.2.5 Juxtaposition analyses

Juxtaposition analyses are specialized moving window functions that compute a summary value based on the position of cells adjacent to the central (focal) cell. Juxtaposition analyses are essential for textural measures (Musick & Grover 1991) and a number of other landscape indices (e.g. Equations 3.8–3.13). They can be also used to create patch borders in raster GIS.

The juxtaposition for a center cell surrounded by four neighbors is given by:

$$J = \Sigma_{n=1}^{4} w_{ij} \qquad (3.14)$$

where w_{ij} are weights assigned to different cell positions relative to the central cell. If X is the central cell, then a weighting matrix of

	8	
1	X	4
	2	

used with Equation 3.14 yields the following results:

0 = no adjacent cells
1 = cell is on left edge
2 = bottom edge
3 = left and bottom edges
4 = right edge
5 = left and right edges
6 = bottom and right edges
7 = left, bottom, and right edges
8 = top edge
9 = top and left edges
10 = top and bottom edges
11 = top, bottom, and left edges
12 = top and right edges
13 = top, right, and left edges
14 = top, right, and bottom edges
15 = cells on all four edges.

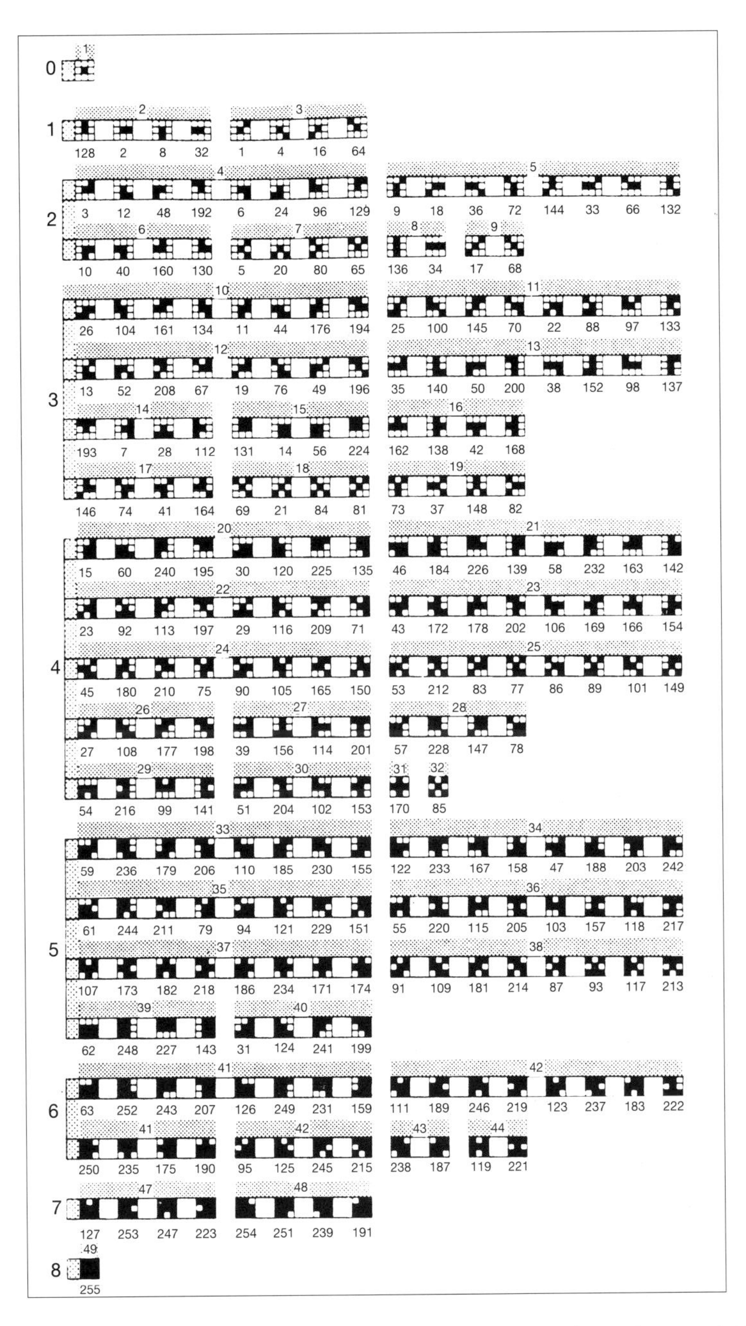

Fig. 3.14 The 256 possible surround conditions for a pixel and its eight adjacent neighbors. Within each of the count groups (labeled 1 to 8 and organized vertically) are subgroups that may contain as many as eight variations consisting of a pattern, its rotations of 90°, 180°, and 270°, and its mirror image, and rotations. Reproduced with permission, the American Society for Photogrammetry and Remote Sensing. Greenlee, (1987). Raster and vector processing for scanned linework. *Photogrammetric Engineering and Remote Sensing,* **53**, 1383–1387.

A similar scheme was developed by Greenlee (1987) for all eight adjacent cells. Use of the following weighting matrix yielded the results in Fig. 3.14:

64	128	1
32	*X*	2
16	8	4

This ability to identify the spatial arrangement of cells is useful in topographic operations (Chapter 4).

The capability of raster GIS to analyze the cells around the edge of a polygon is of special interest to ecologists studying ecotones, zones of transition between adjacent ecological systems (di Castri *et al.* 1988; Risser 1995). Johnston and Bonde (1989) used edge detection software to analyze ecotones within a 324-km^2 Landsat Thematic Mapper image from northern Minnesota, U.S.A. The image was classified into six general land cover categories, processed to remove small patches, and scanned using the boundary option of a moving window GIS routine (see Section 3.2.4). Pixel values along the perimeter of each land cover patch retained their land cover code, while non-boundary pixels were recoded as zero values. The resultant boundaries were two pixels wide, showing the land cover classes present on both sides of each ecotone. This type of analysis is important for analyzing contiguous habitats in wildlife management applications (Mead *et al.* 1981; Heinen & Cross 1983), and can be used to determine if certain plant communities are associated with each other in greater proportion than would be expected from their overall abundance in the landscape as a whole (Pastor & Broschart 1990).

3.2.6 Buffer zones

Most GIS are capable of generating a buffer zone around spatial objects, such as a hazard zone around a chemical spill, or a logging restriction zone adjacent to a stream. These buffers are defined by: (i) the spatial entity for which the buffer is desired; and (ii) a user-specified radius or distance. The width of buffer zones may be uniform or may be varied by entity class. Using the example of our invertebrate species richness data layer (Fig. 3.8), we might exclude intensive land uses from a 100-m buffer strip adjacent to river segments with 45–49 taxa, a 50-m buffer strip adjacent to river segments with 35–44 taxa, and a 25-m buffer strip adjacent to river segments with 34 or less taxa, so as to provide greatest protection for those river segments with the highest biodiversity. Buffer generation requires only a single data layer, but it generates a new data layer that usually is intersected with another (see Section 3.3.3.2).

In raster systems, buffering is done by a spreading process which computes the distance between the feature of interest and every other cell

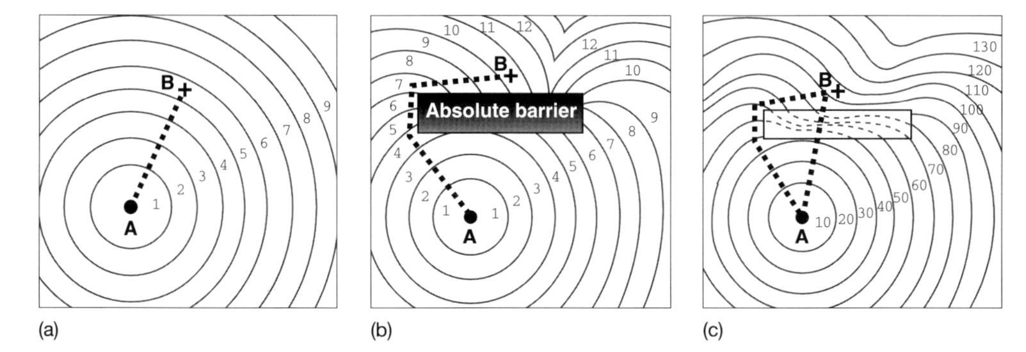

Fig. 3.15 Use of a spreading function in a raster GIS to define 1 km increments around a point of origin: (a) no barriers to movement, (b) effect of an absolute barrier on the spreading function, (c) effect of a partial barrier on the spreading function. Adapted from *Geographic Information Systems: A Management Perspective* by Aronoff (1989). WDL Publications. Ottawa, Canada.

within a defined limit, resulting in halos of cells with incrementally larger distances from the central feature. In some raster GIS packages, additional criteria can be applied to the spread command to control the direction (uphill, downhill), resistance, and barriers to spreading (Fig. 3.15). Such criteria allow the spread process to be used in surface routing, such as the overland flow of surface runoff (see Chapter 4).

3.2.7 Spatial subsetting

Spatial subsetting is most often done in a GIS by intersecting two data layers (Section 3.3.2), but spatial subsetting can also be done in both raster and vector GIS by defining the corners of a rectangle within which to subset data. This type of operation commonly is done to enlarge an area for display purposes or define a map extent for printing.

3.3 Operations performed using multiple data layers

3.3.1 Geometric transformation

A geometric transformation spatially adjusts a data layer so that it can be overlaid correctly on another of the same area. Although the adjustment may be made to only one of the data layers, another layer usually serves as a reference, making it a multiple data layer process. Geometric transformation is an important precursor to database overlay, because spatial mismatch can be a significant source of error.

Maps can have different coordinate systems and projections (Sections 2.4 & 2.5), which must be reconciled before they can be combined. Most GIS are capable of transforming among a number of different projections. However, there can be mismatches even when both data layers have the same coordinate systems and projections. Processes for adjusting such mismatches include registration by absolute position, registration by relative position, conflation, and edge matching.

3.3.1.1 Registration by absolute position

In registration by absolute position, each data layer is transformed independently using ground coordinates, such as those obtained by the Global Positioning System (GPS — see Chapter 8) or conventional surveying. The surveyed reference points should be features such as road intersections and stream confluences that are identifiable on the data layer to be adjusted. The advantage of registration by absolute position is that errors are not propagated from one data layer to another, but the disadvantage is that small positional errors may occur independently in each data layer, so that boundaries which should match precisely may be slightly misaligned (Aronoff 1989). Such misalignment can be corrected by conflation.

3.3.1.2 Registration by relative position

Registration by relative position geometrically transforms a data layer based upon a reference layer that is assumed to be georeferenced accurately (not always a correct assumption). Points are identified that can be located precisely on both data layers, and entered into the GIS with a digitizing table or other data input device. The GIS then calculates the mathematical corrections needed to perform the transformation, expressed as root mean square error (**RMSE**). RMSE is calculated by determining the positional error of the test points, squaring the individual deviations, and taking the square root of their sum. The GIS operator accepts or rejects the transformation based on this statistic. This process of geometric transformation commonly is called "**rubber-sheeting**" because the procedure is analogous to the stretching of a sheet of rubber.

Registration by relative position works best in regions that have been altered anthropogenically, because features such as road intersections and field corners provide easily distinguishable points that can be located precisely on both data layers. This type of registration is more difficult, however, in natural areas that lack anthropogenic alteration.

With this type of registration, positional errors in the reference data layer are propagated to transformed data layers. Therefore, the reference data layer should be the one deemed most accurate, and special care should be taken in identifying and digitizing its reference points.

3.3.1.3 Conflation and edge matching

Conflation is the procedure of reconciling the positions of corresponding features in different data layers (Aronoff 1989). For example, the boundaries of lakes in humid regions are relatively static, and should match on overlaid data layers. Conflation procedures are used to ensure that such features actually do match. The GIS operator identifies corresponding features on both maps, and one or both maps are adjusted.

Edge matching involves a similar concept: lines and polygons that cross map boundaries should match. Although this seems intuitive, minor linework mismatch is common across the edges of adjacent maps and data layers. For example, 1 : 100 000 digital line graphs (DLGs) supplied by the U.S. Geological Survey (USGS) must often be edge matched before use. Special routines for automating or semi-automating edge matching are part of many GIS.

3.3.2 Clipping

Clipping is the spatial subsetting of one data layer using the boundaries of another. Clipping is used to extract information about an area of interest, masking out extraneous material. The data layer used to define the boundaries of the area to be clipped may have been created as a buffer (see Section 3.2.6). Clipping has been used in water quality investigations to determine land uses within buffer zones adjacent to streams. Clip layers may also represent political boundaries (e.g. a county), or other regions of interest. For example, the vegetation cover within a floodplain could be extracted from a regional vegetation map using a clip layer of floodplain boundaries. The converse of clipping is to mask out, rather than clip, the area within the bounds of the defining polygon, retaining information outside its bounds.

3.3.3 Overlay operations

3.3.3.1 Graphical overlay

The GIS function most commonly depicted in popular literature is the overlaying of multiple data layers into a single, comprehensive map (Fig. 3.16), much like the overlaying of line-work drawn on sheets of clear plastic. This type of graphical overlay is standard in vector GIS display and hard copy output, allowing the user to display simultaneously the line-work of two or more superimposed data layers. Many raster GIS can also display line-work in a graphics layer over a raster data layer, such as a DLG file of roads and streams superimposed on a satellite image. This type of overlay is extremely useful, but does not change the characteristics of the component files, nor does it create a new data layer.

In older raster systems, such graphical overlay was not possible without generating a new data layer. Each cell could receive only one classification, so a combination hierarchy had to be specified to perform a graphical overlay. For example, to superimpose roads on a vegetation layer, data for those cells classified as roads on the first data layer would have to replace the data in corresponding cells on the second data layer.

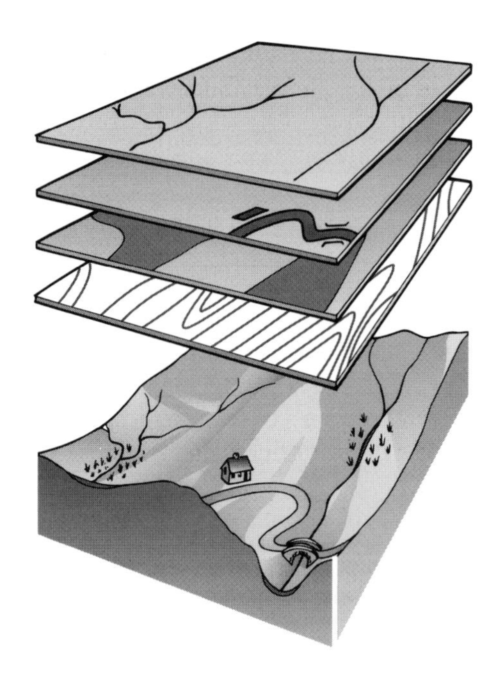

Fig. 3.16 Overlay of multiple data layers. After Burrough 1986, by permission of Oxford University Press.

This replacement could be done by assigning roads a higher classification number than any of the vegetation classes, and overlaying with a maximum replacement algorithm.

3.3.3.2 Logical overlay

Logical overlay employs Boolean operators, previously described relative to database operations (Section 3.1.1.4), to analyze the spatial coincidence of input data layers. In a vector GIS, many new polygons are

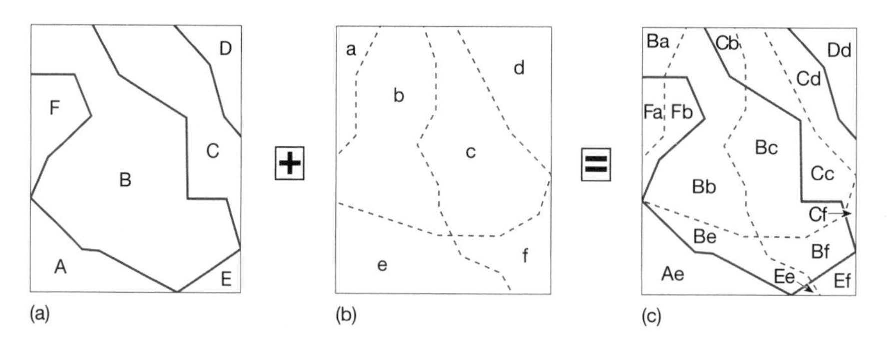

Fig. 3.17 Union of vector data layers. From Burrough 1986, by permission of Oxford University Press.

generated by combining data layers having different internal boundaries, even when the input maps are simple and cover the same area (Fig. 3.17). When the spatial extent of the two input layers is different, different vector overlay procedures yield new maps differing in spatial extent. An **intersect** is used to generate a data layer showing all combinations of data that occur within the area common to both input data layers. A **union** is used to generate a data layer showing all combinations of data that occur within the extent of either data layer. An "XOR" combination (called "**identity**" in the ARC/INFO GIS) is used to generate a data layer showing all combinations of data that occur within the spatial extent of the first data layer. Logical overlays are done usually with polygon data layers, but intersect and identity overlays can also be done between a polygon data layer and a point or line data layer.

In a raster GIS, Boolean expressions can be used to perform spatial queries of two or more files, yielding a new data layer with the results. For example, the statement:

If "clay bluff" AND "southwest facing", then "erosion = low"

could be used with data layers of lakeshore geology and aspect to classify the shoreline erosion hazard of southwest-facing clay bluffs.

Boolean operators can also be used in combination. The statement:

If "bedrock bluff" OR ("clay bluff" AND "southwest facing"), then "erosion = low"

indicates that erosion is low for all bedrock bluffs and for southwest-facing clay bluffs. Boolean operations in raster GIS may be performed using any number of data layers and conditional statements.

An intersect procedure can be used in a raster GIS to determine the spatial coincidence of different class combinations between two data layers. The properties of a location in one data layer are compared with the properties of the same location in another data layer and a new data layer is generated based on the results, with class numbers assigned according to a non-commutative matrix. This capability is useful in change detection (see Section 6.4.1).

Logical overlays are often used in ecology to derive empirical relationships for plant and animal habitat. For example, the home ranges of several radio-collared moose could be compared to locate areas where intra-specific competition for resources is greatest. Boolean operations can also be used with range distribution maps for multiple plant or animal species to optimize areas for nature preserve siting. Boolean analysis can locate areas having unique combinations of habitat variables required by plants or animals, or when these habitat variables are unknown or poorly understood, Boolean operations can be used to deter-

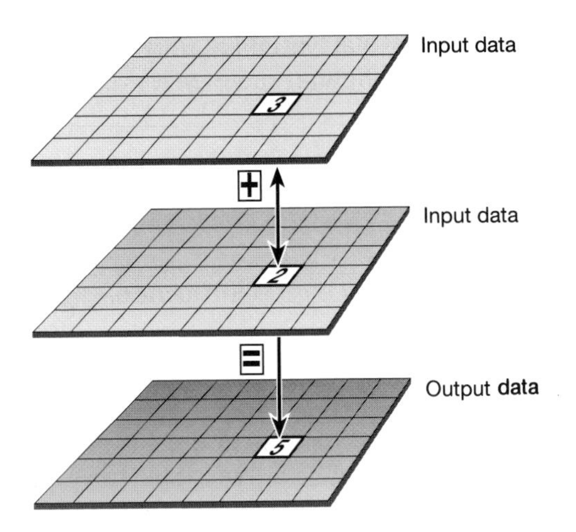

Fig. 3.18 Arithmetic overlay of two raster data layers using an addition operation. Adapted from *Geographic Information Systems: A Management Perspective* by Aronoff (1989). Published by WDL Publications. Ottawa, Canada.

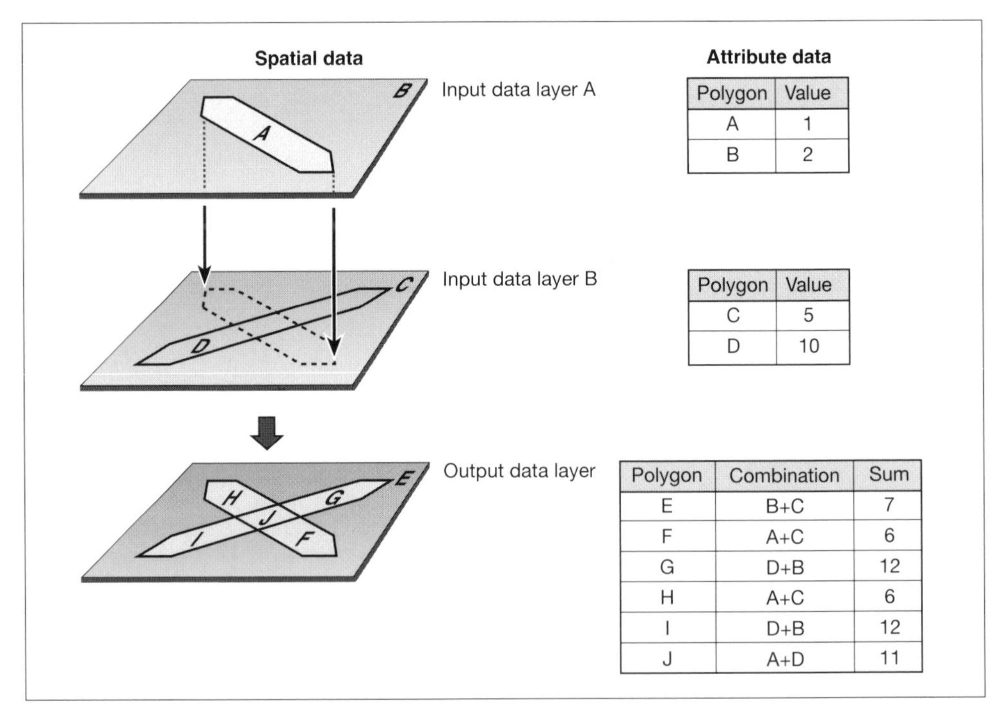

Polygon	Value
A	1
B	2

Polygon	Value
C	5
D	10

Polygon	Combination	Sum
E	B+C	7
F	A+C	6
G	D+B	12
H	A+C	6
I	D+B	12
J	A+D	11

Fig. 3.19 Arithmetic overlay of two vector data layers using an addition operation. Adapted from *Geographic Information Systems: A Management Perspective* by Aronoff (1989). Published by WDL Publications. Ottawa, Canada.

mine them by intersecting habitat variables with organism distribution. In this way, a GIS can identify habitat preferences which may be difficult to discern using other methods.

3.3.3.3 Arithmetic overlay

Overlay operations can also be used to perform mathematical operations across multiple data layers (Fig. 3.18). For example, the Universal Soil Loss Equation developed by the U.S. Department of Agriculture is computed as:

$$A = RKLSCP \tag{3.15}$$

where A is annual soil loss, R is the rainfall erosion index, K is inherent soil erodibility, LS is the combination of slope percentage and slope length, C is the cover and management factor, and P is the conservation practice factor (Wischmeier & Smith 1978). Soil loss can be computed for each cell in a raster data layer by cross-multiplying raster data layers representing each of the variables in the equation.

In a vector GIS, these input variables would be derived from data layers of land cover/management and soil characteristics. Because the boundaries of these features would not necessarily match, it would first be necessary to use a logical overlay to generate new polygons, as in Fig. 3.17. The numerical values associated with those polygons could then be multiplied as a database operation (Fig. 3.19).

CHAPTER 4

Topographic operations

Topography is an important driving variable in many ecological processes because of its influence on insolation, water flow, and organism movement. Topographic maps are widely available, and digital representations of elevation are becoming increasingly available. GIS provides a number of methods for analyzing topography, many of which can be meaningfully applied to any continuous data layer (defined in Section 2.2.3.3). The ability to analyze digital topographic data has significantly advanced ecological and hydrological modeling.

4.1 Elevation data

Topographic maps generally classify elevation data into intervals and represent those intervals by isolines (a vector representation) known as contour intervals. This is a perfectly acceptable way to represent three-dimensional features on a two-dimensional surface, and most map users have a good understanding of the topographic reality depicted by this representation: a contour line indicates constant elevation, concentric circles indicate topographic hills or depressions, the spacing between lines is an indication of slope, and the "Vs" that occur where contour lines intersect streams always point upstream.

Although this representation works well on paper maps, there is no reason why elevation data should be stored in this form in a GIS. Just as latitude and longitude coordinates are continuous across the surface of the Earth, elevation is also continuous. It is impossible to record the elevation of every point on the Earth's surface, just as it is impossible to record the latitude and longitude coordinates for every point on the Earth's surface. Therefore, recorded elevation data are of necessity a sample.

4.1.1 Where do elevation data come from?

Digital elevation models (**DEMs**), also known as digital terrain models (DTMs), are used to represent altitude data in a GIS. Most map users do not concern themselves with the source of the elevation data represented by a topographic map or DEM, but as with any data set, an understanding of the source data is important to its interpretation. Any topographic data set starts with **ground control** points of known location (horizontal control) and elevation (vertical control). Ground control has traditionally been measured by surveyors, but the development of Global Positioning Systems (GPS — see Chapter 8) has made it easier for others to obtain accurate ground control. With a sufficient number of ground control points, a topographic map can be constructed by interpolation (see Chapter 7). However, because traditional surveying techniques are very time-consuming, this method is generally restricted to small areas.

Stereoscopic plotting instruments (**stereoplotters**) have been the primary means of generating elevation data for topographic maps during the twentieth century. A stereoplotter uses a pair of aerial photographs taken at successive positions along a flight path, such that there is about 60% overlap in the areas covered, called a **stereopair**. Each photograph in a stereopair is the result of rays projected from the terrain, through the lens of the aerial photograph camera, onto the image plane of the film. Because the camera was in two different locations when each of these photographs was taken, the angles of these rays change from photograph to photograph, and the relative position of stationary objects appears to change due to stereoscopic displacement (Fig. 4.1). When viewed simul-

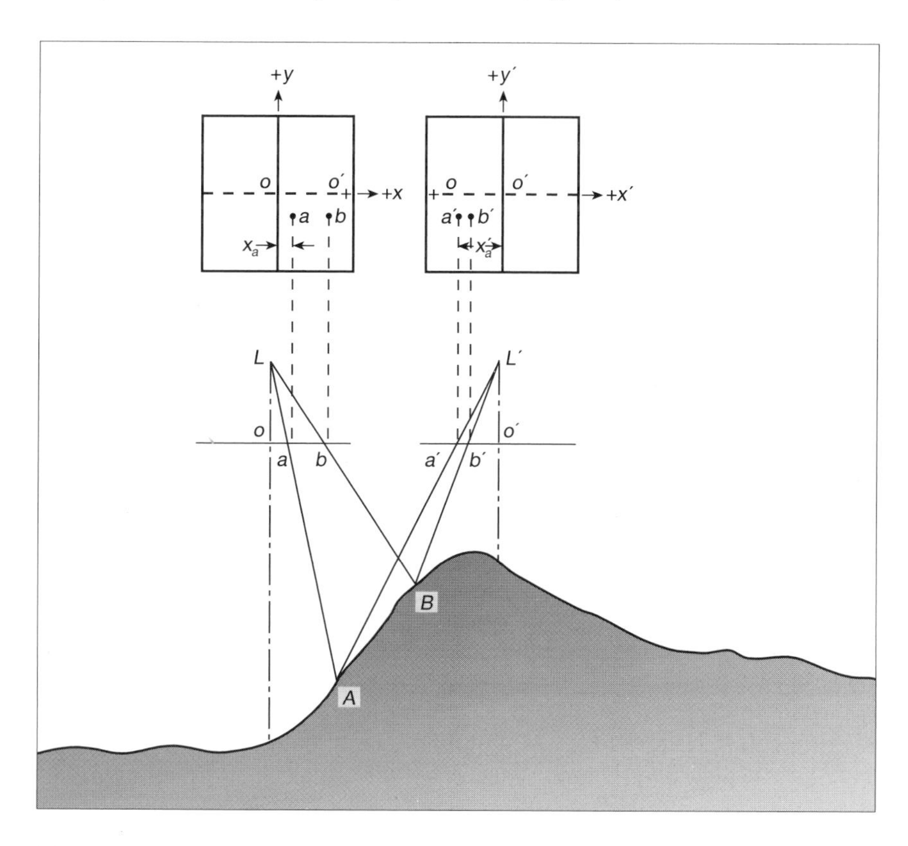

Fig. 4.1 Parallax displacements on overlapping vertical photographs. From *Remote Sensing and Image Interpretation* by Lillesand & Kiefer, © 1994. Reprinted by permission of John Wiley & Sons, Inc.

taneously, a stereopair portrays a three-dimensional image, which is essentially a reduced scale model of the terrain in the overlap area. To generate contours, the stereomodel is projected onto a tracing table platen, which is set at a particular elevation, and the stereoplotter operator generates contours by moving a floating mark along the terrain so as to maintain contact with the surface of the model (Lillesand & Kiefer 1994). Thus, the data generated by a stereoplotter are in the form of contour lines tied to elevation by ground control points.

Digital elevation is currently obtained as a by-product of generating **orthophotographs**, aerial photographs in which image displacement due to photographic tilt and relief has been substantially eliminated. Like the process of stereoplotting, orthophotograph production uses stereopairs of aerial photographs. The stereomodel is scanned in closely spaced transects using a floating mark, and these scan line elevation profiles are stored in digital form (Fig. 4.2a). The elevation profiles are then read by instruments that automatically raise and lower a film holder to expose the orthophotograph negative.

When digital elevation data are in the form of contours, either scanned from existing contour maps (USGS 1987) or produced directly in digital form by a stereoplotter, data density is greatest along the elevations chosen for contours and lowest in level terrain at elevations lying between contours (Fig. 4.2b). Interpolation techniques can be used to increase the density of the data between contours, but even when this is done, there tends to be a periodicity in the data that is an artifact of its origin. Figure 4.3 shows a frequency plot for elevation data obtained from a 7½′, 30 × 30 m grid DEM produced by the U.S. Geological Survey (USGS). The wave pattern that is discernable in this plot is due to the fact that there are more data points at elevations that were contours on the source map than there are in interpolated areas between them.

Digital elevation data obtained from profiles, as in the case of DEMs generated as a by-product of orthophotograph production, tend to be more evenly distributed (Fig. 4.2a). Data density is greater in steeper areas than in level areas, but tends to be less spatially aggregated than data derived from contours (Carter 1988).

Inflection points in a landscape (e.g. stream courses, ridge lines, peaks, low points, and breaks in slope) are topographically significant, but are not well-represented by digitized contour lines or scan line elevation profiles (Fig. 4.2). Methods for including inflection points in digital data have greatly improved the accuracy of a DEM, making them much more useful for flow-routing applications (Hutchinson 1993).

4.1.2 Topographic representation in GIS

Regardless of the method used to gather the original elevation data, DEMs are almost always provided in grid cell format by the national

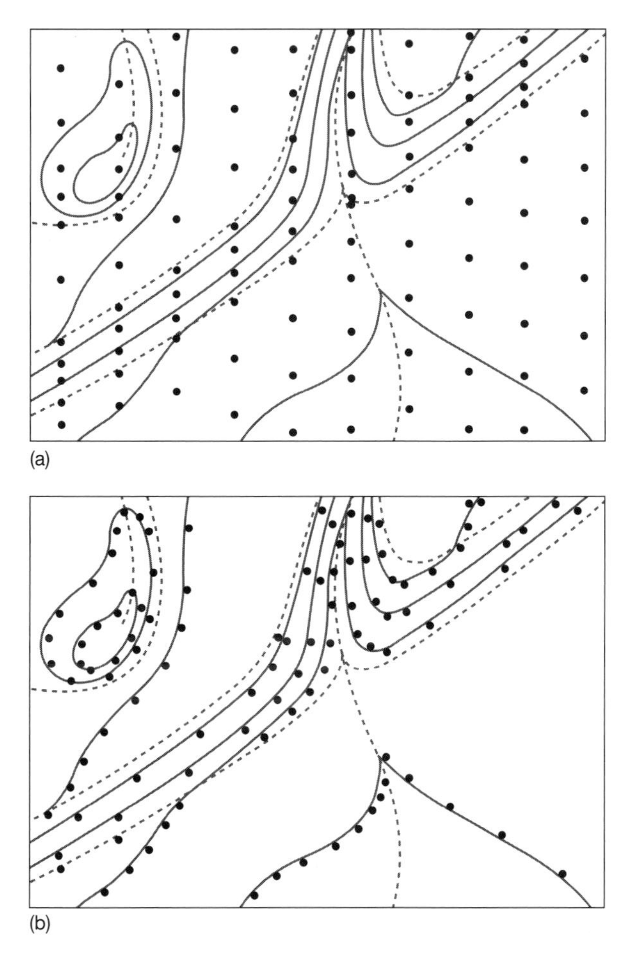

Fig. 4.2 Two forms of DEM data capture: (a) scan line elevation profiles, (b) digitized contours. Reproduced with permission, the American Society for Photogrammetry and Remote Sensing. Carter, (1988). Digital representations of topographic surfaces. *Photogrammetric Engineering and Remote Sensing*, **54**, 1577–1580.

governments that compile and distribute them. The U.S., U.K., and Australia all have nationwide coverage by coarse DEMs, with cell sizes ranging from about 90 × 90 m in the U.S. to about 2500 × 2500 m in Australia (Smith *et al.* 1989; Hutchinson 1993; Steyaert 1996). DEMs derived from medium-scale maps (~1 : 25 000) that have finer grid cell resolution (e.g. 30 × 30 m) are becoming increasingly available in the U.S. and the U.K. (Miller 1996; Steyaert 1996). Once obtained, grid cell DEMs can be used as they are, or converted into other forms of representation.

In vector GIS, digital elevation is generally converted into a Triangulated Irregular Network, or **TIN** (Peucker *et al.* 1078). A TIN is a terrain

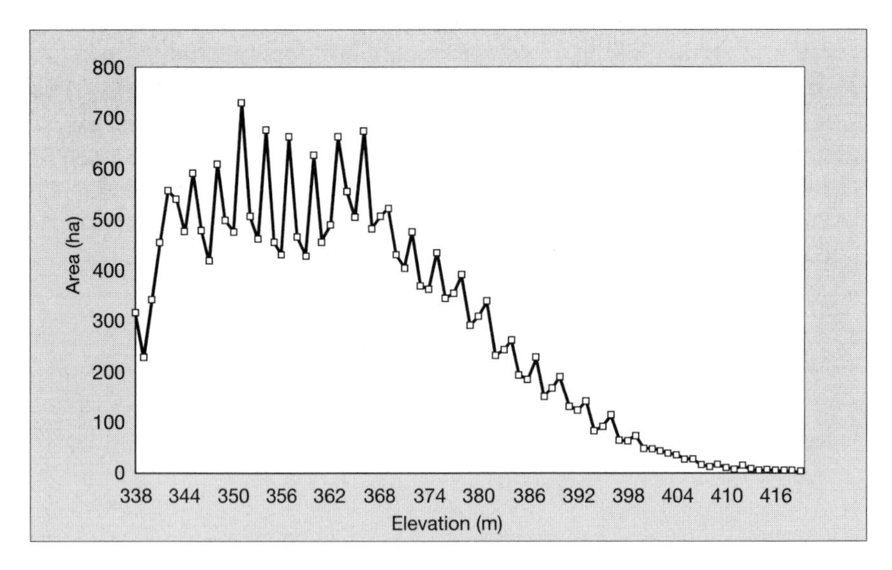

Fig. 4.3 Frequency plot for elevation data obtained from a 1 : 24 000 DEM produced by the USGS for Voyageurs National Park in Minnesota.

model in which the land surface is covered by continuous, connected triangles, representing planar facets of the landscape (Fig. 4.4). TIN triangles are large in flat (although not necessarily level) portions of the landscape and small in areas of complex terrain. A TIN is a more efficient means of data storage than a raster DEM: a well-constructed TIN of 100 points can represent a data surface as comparably well as a DEM of several hundred points. A TIN also works well in areas with sharp breaks in slope, such as long ridges or channels, because triangle edges can be aligned with breaks. A TIN representation is particularly suitable for depicting fluvially eroded landscapes, but is less suitable for glaciated landscapes.

Most GIS software packages can generate elevation surfaces from point data input (see Chapter 7) and can generate contour plots as graphical output. However, topographic analysis generally is done with grid cells (raster GIS) or TINs (vector GIS).

4.2 Topographic analysis

Topographic operations analyze the change in attributes over space. These operations are most often applied to elevation data, but pertain to analysis of any continuous data surface. It is easy to visualize how these operations could be used with depth instead of elevation data (e.g. lake water depth to analyze bathymetry, ground water depth to analyze the water table surface), but they could just as easily be applied to any spatially distributed continuous data surface, such as atmospheric pressure or estuarine water quality. In the latter case, the *z*-value no longer

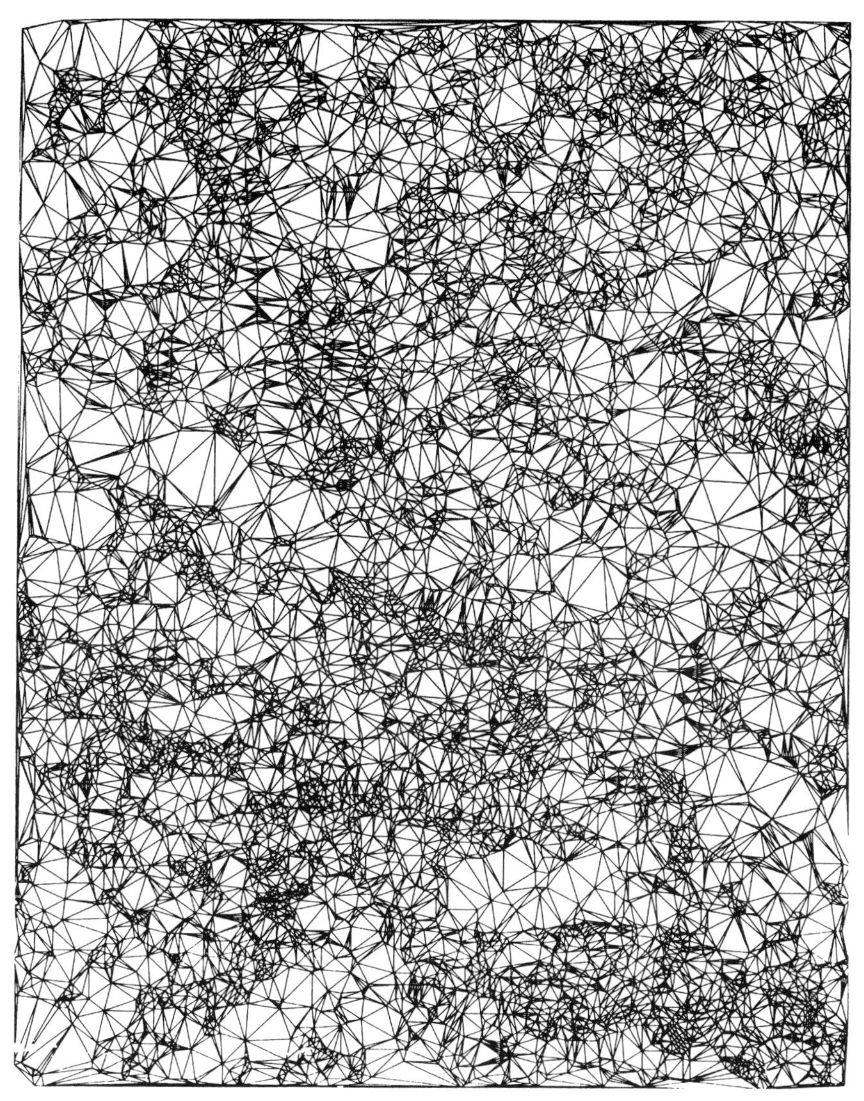

Fig. 4.4 A TIN model of elevation. Reproduced with permission, the American Society for Photogrammetry and Remote Sensing. Pike *et al.* (1987). A topographic base for GIS from automated TINs and image-processed DEMs. In *GIS '87 Proceedings*, 340–351.

represents the third spatial dimension, but instead represents an attribute value measured at an x,y location.

4.2.1 Surface-specific points

Certain points on a topographic surface occur at key topographic structures, such as peaks, pits, ridges, and channels. Not only are these surface-specific points topographically significant, they are also an important part of a well-constructed TIN. They can be located by simple algorithms that analyze the spatial distribution of elevation differences between a focal cell and its eight neighbors within a 3 × 3 moving window (Fowler & Little 1979; Goodchild & Kemp 1990). A point is a **peak** if its eight neighbors are all lower, and it is a **pit** (sink) if its eight neighbors are all higher. A point is a **pass** if higher (+) and lower (–) points alternate around the point with at least two complete cycles:

```
+   +   –            +   –   –
–       –     OR     –       –
–   +   +            +   –   +
(2 cycles)           (4 cycles)
```

Next the surface is examined using a moving 2 × 2 window, such that each point (except those along the edges of the data layer) occurs in four different window positions. If a point does not have the lowest elevation in any of the four window positions, then it is a potential **ridge** point. If a point does not have the highest elevation in any of the four window positions, then it is a potential **channel** point. Then starting at a pass, the adjacent ridge points are searched until a peak is reached, and the adjacent channel points are searched until a pit is reached. The result is a connected set of peaks, pits, passes, ridge lines, and channel lines (Goodchild & Kemp 1990).

4.2.2 Slope

As used here, **slope** (gradient) is the rate of change in elevation. In the terminology of calculus, slope can be thought of as the first derivative of elevation, differentiating a continuous surface to generate a gradient map as output. Some authors distinguish slope from gradient (Evans 1980), but the two terms are used here interchangeably.

Hill slopes are measured by field scientists with a clinometer, which gives the slope of a linear transect between two points as an angle relative to horizontal, or as a percentage ratio of vertical to horizontal distance. Linear slopes are calculated on a topographic map as the "rise over the run," or the change in elevation divided by the distance between two points:

$$S = \sqrt{(\delta Z/\delta X)^2 + (\delta Z/\delta Y)^2} \tag{4.1}$$

where δZ is change in elevation and δX and δY are the change in horizontal coordinates (see Equation 3.1, the Cartesian formula for computing distance). Taking the tangent of S provides slope in degrees; multiplying S by 100 expresses it as a percentage.

In a raster GIS, slope can be computed using the linear slope values from the focal cell in a 3 × 3 moving window to the eight surrounding

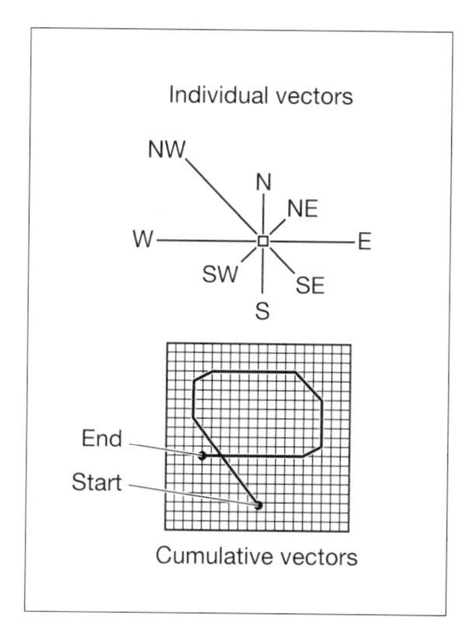

Fig. 4.5 Determining slope for a 3 × 3 kernel, Cumulative vectors. By permission. *Beyond Mapping: Concepts, Algorithms, and Issues in GIS*. Berry. © 1996, GIS World, Inc.

cells. The **distance-weighted drop** is calculated by subtracting the neighbor's elevation value from the center cell's elevation value and dividing by the distance between the cells. Then different algorithms choose one or more of the values to compute the slope value for the focal cell. For example, slope can be determined as: (i) the gradient of the steepest distance-weighted drop to one of the eight neighboring cells (called the deterministic 8-node algorithm); (ii) the gradient of the steepest drop OR the steepest rise from the center cell, whichever is greater; or (iii) the average gradient of the eight nearest cells.

Surficial slope is the angle at which a surface is inclined relative to a horizontal plane. As with linear slope, surficial slope can be expressed in degrees of arc or as a percentage. Surficial slope is easily calculated in a TIN as the angle between the plane of each triangle and the horizontal plane. Slope can be calculated in the same manner with a grid cell DEM, but first trend surface analysis must be used to fit a plane to the points surrounding a focal cell in a 3 × 3 moving window, so as to minimize the sum of the squared elevation differences between the plane and the data points.

Surficial slopes can also be determined by means of vector algebra as follows

1 Individual vectors the relative length of the slope value are drawn in the direction of each of the eight adjacent cells.

2 Starting with the longest vector, successively connect each of the vectors, working in a clockwise direction.

3 A vector drawn between the starting and ending point (resultant vector) has a length representing the slope of the plane, and a direction representing its aspect.

For display purposes, the raw slope values computed in a raster GIS are often simplified by spatial aggregation into larger cells or by reclassification into larger data intervals (see Chapter 3).

4.2.3 Aspect

Slope azimuth or **aspect** is the direction which a surface faces, usually expressed in degrees (0–360°) or as compass headings (N, NE, E, SE, S, SW, W, NW). Aspect is calculated in a TIN as the direction of maximum slope of the plane defined by each triangle. If a plane has been fitted to a 3 × 3 window of grid cells to compute slope, then aspect is the direction of maximum slope of that plane.

Where slope has been determined in a raster GIS as the cell with the steepest drop, the aspect is the direction of that cell. Conversely, where slope has been determined to be the cell with the steepest rise from the center cell, the aspect is the direction of the opposite cell. The aspect of a slope determined by vector algebra is the direction of the resultant vector.

Aspect is an important characteristic of hillslopes because of its effect on solar illumination. Aspect is also an important characteristic of coastlines, because of its relationship to prevailing wind direction, which can affect shoreline erosion (Johnston & Salés 1994; Johnson & Johnston 1995). The aspect of a coastline is the direction perpendicular to the line segments that compose it or to a line that is fit to a series of points along the coast (see Section 7.2.1).

4.2.4 Inflection

Just as slope is the first derivative of elevation, inflection can be thought of as its second derivative or the rate of change of slope. Inflection, also known as the **profile curvature**, is the curvature of a surface in the direction of slope. Areas of low inflection occur where slopes are uniformly planar within the area defined by the analysis (e.g. the dimensions of a 3 × 3 window), whereas areas of high inflection occur where slopes change rapidly over space, such as the crest of a hill or a break in slope from a steep to a level area. Note that a steeply sloped surface has low inflection if its surface is planar. In a TIN, areas of low inflection tend to be represented by larger triangles than areas of high inflection, although this is not necessarily so because triangle size is also a function of the user-selected points chosen to represent the surface. Inflection is measured in degrees of slope per unit of distance (e.g. degrees per 100 m).

The curvature of a surface perpendicular to the direction of slope is its **planform curvature**. Planform curvature indicates where the surface is concave or convex (ESRI 1992). These slope shapes are important in

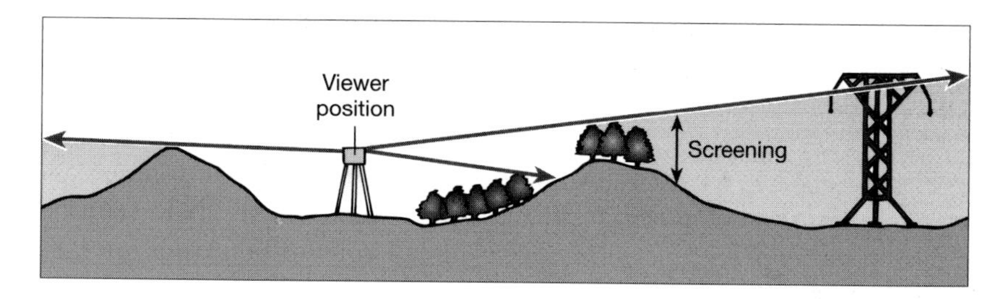

Fig. 4.6 The concept of viewshed. Adapted from *Geographic Information Systems: A Management Perspective* by Aronoff. WDL Publications. Ottawa, Canada.

modeling waterborne material tranport, because erosion tends to occur on convex slopes, whereas deposition tends to occur on concave slopes.

4.2.5 Line-of-sight maps

A GIS can be used to generate maps identifying the intervisibility of landscape features. A **viewshed** shows the land area that can be seen from a specified observation point. Conversely, a **visual impact** map shows all observation points from which a tall object (e.g. building transmission tower) can be seen. The analysis is essentially the same for both operations, the difference being that there is a single observation point for viewing multiple targets in a viewshed, versus multiple observation points for viewing a single target in a visual impact map. In ecology, a viewshed analysis might be used to depict the terrain visible from a raptor's nest, and a visual impact map might be used to determine the visibility of a planned transmission tower from a wilderness area.

Input data layers include elevation, height and location of the observation point (viewshed map) or target (visual impact map), height and location of features projecting above the land surface that could block the view (e.g. buildings, trees), and optionally the maximum viewable distance (Fig. 4.6). The view may be omnidirectional or it may be limited by vertical and horizontal field-of-view specifications. The output is a data layer with a binary classification. For a viewshed, the output is a data layer classified into those areas that are visible and those that are not; for a visual impact map, the output is a data layer classified into those areas that can see the target object and those that cannot.

4.2.6 Illumination

Illumination operations calculate the **reflectance** of an elevation file based on the slope and aspect of land surface facets and the angle and azimuth of the sun. Reflectance values are usually expressed as the sine of the angle between the slope and the sun, and can be displayed as a shaded relief map, which uses light and shade to portray three-dimensional objects (Fig. 4.7). For aesthetic reasons, the light source is usually chosen

Fig. 4.7 Shaded relief map generated from a DEM. Reproduced with permission, the American Society for Photogrammetry and Remote Sensing. Pike *et al.* (1987). A topographic base for GIS from automated TINs and image-processed DEMs. In: *GIS '87 Proceedings*, 340–351.

as being at an angle of 45° above the horizon to the northwest, "a position that has much more to do with human facilities for perception than it does with astronomical reality" (Burrough 1986).

A related GIS capability that is much more useful in ecological applications is the calculation of **insolation**, the amount of clear-sky solar radiation received by a land surface. Insolation is a function of elevation, slope, aspect, latitude, time of year, and shading by other portions of the landscape (Dubayah & Rich 1996). Insolation is an important environmental variable due to its relationships with plant growth, soil moisture, and snow melt.

4.2.7 Surface water flow

Water flows downhill. This simple concept underlies surface water hydrologic modeling. Rain falling upon the land surface flows downslope in a direction that can be predicted from elevation data, and combines with the flow from other portions of the landscape. When sufficient flow accumulates, a channelized stream forms. Although simple in concept, the implementation of this principle in a GIS provides hydrologists and ecosystem scientists with a powerful tool for simulating the location and quantity of surface water runoff.

Land surface flow modeling is a rich and growing field of GIS application, which can only be touched upon here. Excellent reviews of literature and developments in this field have been written by Maidment (1993, 1996), Moore *et al.* (1993), and Wilson (1996).

4.2.7.1 Flow routing with grid cells

Jensen and Domingue (1988) developed a comprehensive procedure for deriving stream flow networks from DEMs, building upon earlier work by O'Callaghan and Mark (1984), Marks *et al.* (1984), and others. Three data layers, all produced from a DEM, are required for this method: (i) a DEM with depressions filled; (ii) a data layer indicating the flow direction for each cell; and (iii) a flow accumulation data layer in which each cell receives a value equal to the total number of cells that drain into it. This method has been incorporated into hydrologic modeling tools developed for ARC/INFO GRID (Fig. 4.8), but similar programs have been developed for other raster GIS (e.g. r.flow for GRASS).

The first step in the procedure is to generate a depressionless DEM. DEMs almost always contain depressions that block flow routing (i.e. water flow stops when reaching the depression). Such depressions are usually data artifacts rather than topographic reality. This step raises cells to the lowest elevation on the rim of the depression, such that each cell in the depressionless DEM is part of at least one monotonically decreasing path of cells leading to an edge of the data layer. The seven-step method for filling depressions in a DEM is detailed in Jensen and Domingue (1988).

The second step in the procedure is to define flow direction for each cell by choosing the direction of steepest distance-weighted drop from among the eight neighboring cells (deterministic 8-node algorithm – see Section 4.2.2). Four sets of conditions can arise: (i) the cell is a pit (see Section 4.2.1); (ii) the distance-weighted drop is larger for one cell than the other seven; (iii) two or more cells are equal in having the greatest distance-weighted drop; (iv) the cell is located in a level area and the direction to the outflow point is not known. Condition (i) is eliminated by generating a depressionless DEM. Condition (ii) yields a clear-cut flow

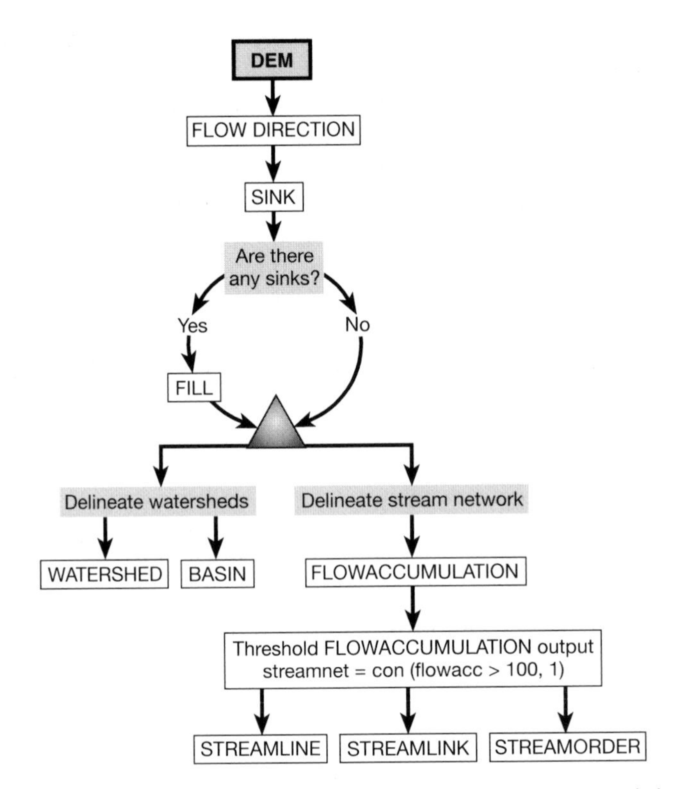

Fig. 4.8 Outline of steps to derive surface characteristics from a DEM. Graphic image supplied courtesy of Environmental Systems Research Institute, Inc. © 1992.

direction, and no further analysis is needed. Condition (iii) can be solved by developing simple rules in a look-up table (e.g. "if three adjacent cells along one edge of the neighborhood have equal drops, assign the flow direction to the middle cell of the three"). Condition (iv) is more difficult to evaluate, and requires an interactive process in order to make flow direction decisions, described in Jensen and Domingue (1988). The chosen flow direction is encoded according to the scheme:

64	128	1
32	*X*	2
16	8	4

such that if cell *X* flows to the left in the matrix, its flow direction will be encoded as 32*. The surround conditions depicted in Fig. 3.14 can also

*ARC/INFO users should note that the numbering scheme in this matrix is rotated clockwise one cell in the FLOWDIRECTION function in GRID, such that the flow direction just described would be encoded as 16.

be temporarily encoded when the flow direction is ambiguous (conditions (iii) and (iv)).

The third step in the procedure is to create the flow accumulation data layer, where each cell is assigned a value equal to the number of cells that flow into it (O'Callaghan & Mark 1984). Cells with a high flow accumulation are areas of concentrated flow, and may be used to identify stream channels. Cells with a flow accumulation of zero are local topographic highs. Flow accumulation values can be multiplied by a grid representing the spatial distribution of rainfall during a storm to calculate rainfall amounts that potentially could be received by each cell, assuming no interception, evaporation, or infiltration (ESRI) 1992). Flow accumulation data layers can be used with Greenlee's (1987) raster-to-vector conversion process to produce a drainage network data layer (Fig. 4.9).

Although the Jensen and Domingue (1988) procedure is now widely used, ongoing research is leading to further developments. Moore (1996) evaluated the basic algorithm used to determine flow direction, and concluded that the deterministic 8-node (D8) algorithm used by Jensen and Domingue (1988) has significant deficiencies, including the inability to model flow dispersion and the production of long, linear flow paths

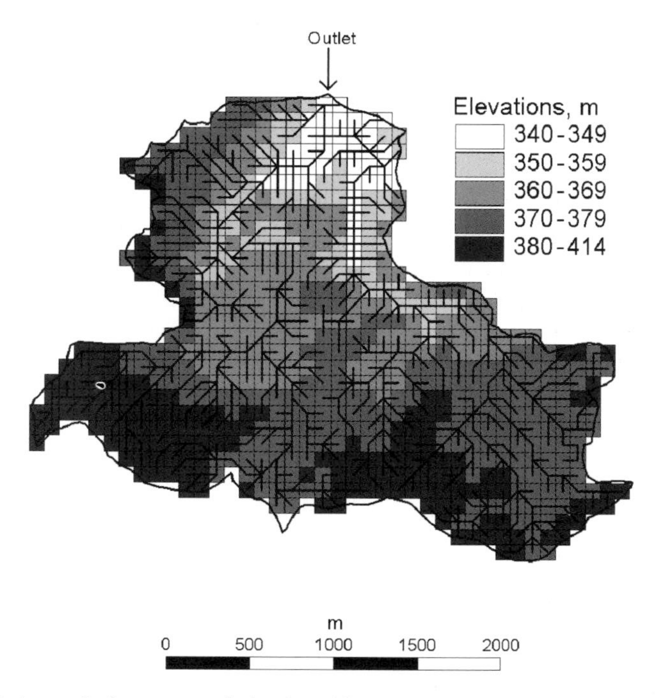

Fig. 4.9 A stream drainage network developed from a DEM-derived flow accumulation data layer. From Nawrocki *et al.* 1994.

that appear to be unrealistic in many landscapes. Mitasova and co-workers (1996) developed an alternative method of computing the geometry of surfaces based on *d*-dimensional differential geometry. In addition to generating scalar fields of slope and aspect, this methodology computes flow path lengths and upslope contributing areas, and uses a flow-tracing algorithm to generate flow lines defining channels and ridges (Fig. 4.10). The Mitasova *et al.* (1996) method has been incorporated into the r.flow procedure for GRASS.

4.2.7.2 Flow routing with other data structures

Grid-based DEMs generally provide the most efficient structures for estimating terrain attributes (Moore *et al.* 1993), so flow routing methodologies for TINs and other data structures are less common. The irregularity of the TIN makes attribute computation more difficult than for the grid-based methods, and the upslope connection of a TIN facet can be difficult to determine. However, the triangular mesh is a common basis for the finite element solution of flow and transport problems, suggesting

Fig. 4.10 Geometrical parameters of land surfaces using *d*-dimensional topographic analysis: (a) flow lines generated downslope, merging in valleys, (b) flow lines generated upslope, merging on ridges. From Mitasova *et al.* 1996. By permission. *GIS and Environmental Modeling: Progress and Research Issues*. Goodchild *et al.* © 1996, GIS World, Inc.

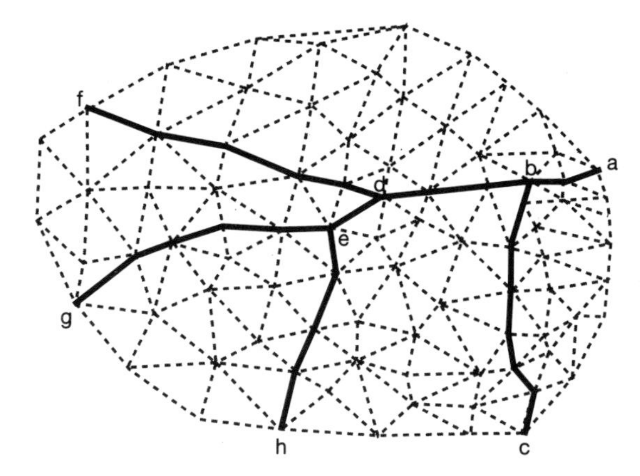

Fig. 4.11 A TIN model of a watershed and its stream system. From Jones *et al.* 1990. By permission of the ASCE.

a linkage between the TIN data structure in GIS and the triangular mesh in finite element codes (Maidment 1993). In finite element modeling, the dependent variables in the governing equations of motion, such as discharge and water surface elevation, are determined at each node in the triangular mesh, and their variation between nodal points is approximated by basic functions (usually linear or quadratic equations), so that a surface of the dependent variable can be constructed over the (x,y) plane. TINs have also been used for stream network and watershed delineation (Fig. 4.11).

Moore (1996) used a digital contour line data layer to divide the 17-km^2 Coweeta watershed, a Long Term Ecological Research (LTER) site in North Carolina, into a series of interconnected hydrologic elements (Fig. 4.12), and modeled flow using a stream-tube analogy. Elements are formed by adjacent contour lines and a pair of adjacent streamlines that are orthogonal to the contours. Catchment areas are determined by accumulating element areas down a stream-tube, and flow direction is computed as the orthogonal to the contour line, in the downstream direction.

4.2.7.3 Watershed delineation

A watershed, also called catchment or drainage basin, is the area of land that contributes precipitation runoff to a waterbody. Watersheds are normally defined to a point on a stream (pour point), but also surround lakes and depressions (Fig. 4.13). Before the advent of GIS, watershed boundaries were determined by visually locating and delineating the maximum elevation separating adjacent drainage basins on topographic maps. In a GIS, the same process can now be done automatically (Marks *et al.* 1984; Band 1986).

Fig. 4.12 Discretization of the Coweeta catchment into a series of interconnected elements using the TAPES-C method of analysis. From Moore 1996. By permission. *GIS and Environmental Modeling: Progress and Research Issues*. Goodchild *et al.* © 1996, GIS World, Inc.

Delineation of specific watersheds is done with two data layers: (i) a flow direction data layer; and (ii) a data layer containing cells or groups of cells that represent the outflow points of the desired watersheds (Jensen & Domingue 1988). Outflow points may be a specific site along a stream where a monitoring station occurs, a block of cells spanning a stream to represent the width and height of a dam, or the cells in a depression such as a pothole. Subwatersheds of a larger drainage basin can also be automatically delineated by defining an area threshold to constrain their minimum area.

4.3 Topographic visualization

GIS typically depict spatially distributed data as they would be shown on

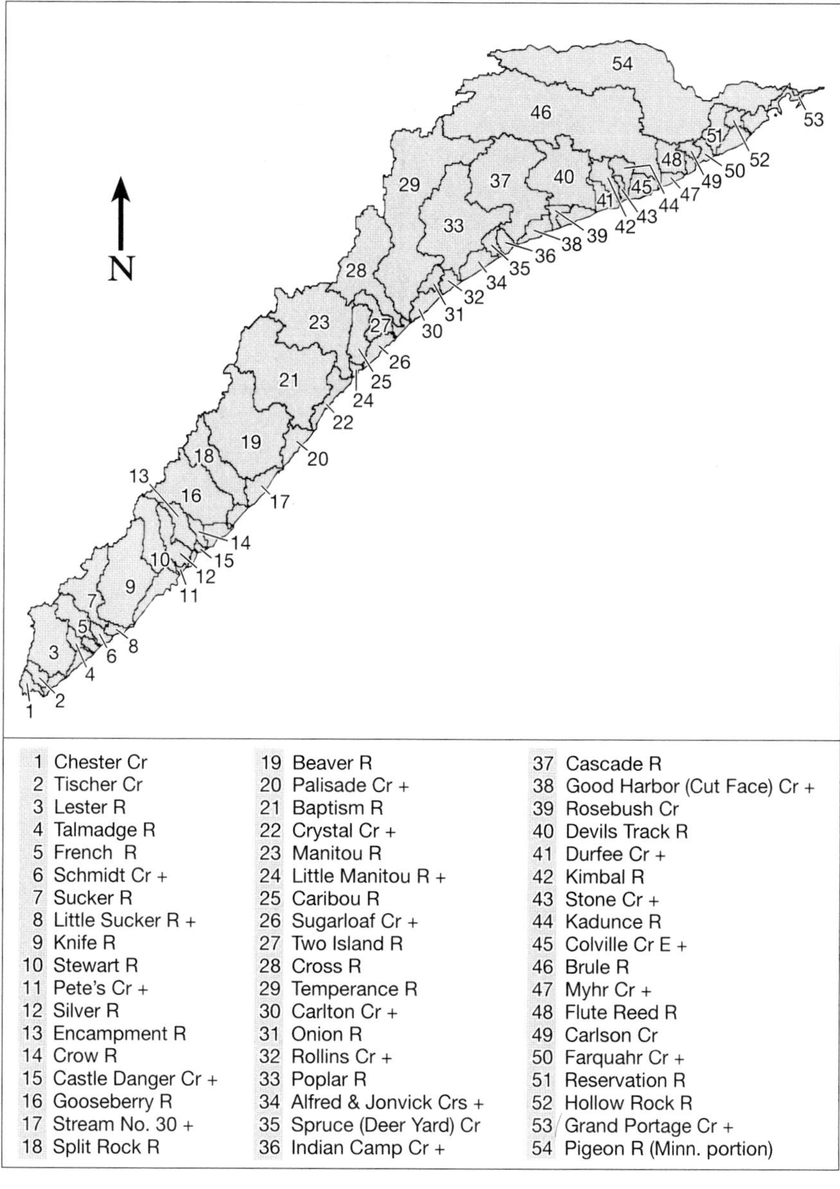

Fig. 4.13 Major watersheds of the Minnesota north shore of Lake Superior (+ indicates areas containing drainage directly to the shore as well as to named creek). From Johnston *et al.* 1991.

a map: two-dimensional surfaces viewed from nadir via a high platform, with spatial objects represented by lines and patterns. The contour map is the most familiar two-dimensional depiction of a three-dimensional

(a)

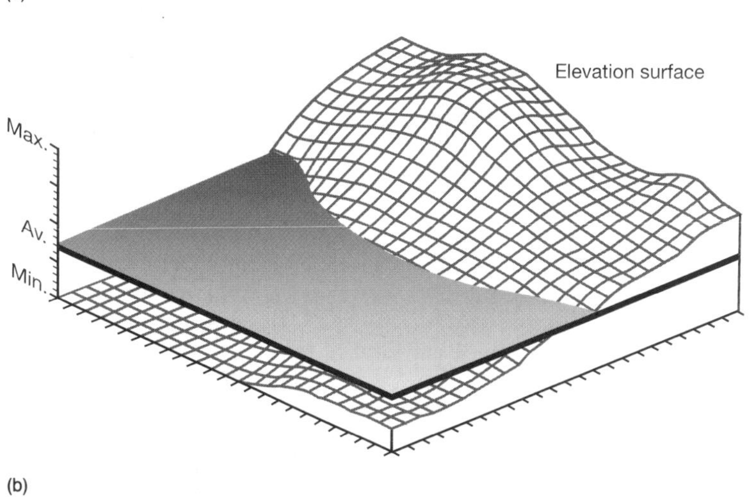

(b)

Fig. 4.14 Perspective views of topographic surfaces: (a) Closely-spaced parallel lines. Reproduced with permission, the American Society for Photogrammetry and Remote Sensing. Pike *et al.* (1987). A topographic base for GIS from automated TINs and image-processed DEMs. In: *GIS '87 Proceedings*, 340–351. (b) Grid cell mesh. By permission. *Beyond Mapping: Concepts, Algorithms, and Issues in GIS*. Berry. © 1993, GIS World, Inc.

surface, but other examples include TINs (Fig. 4.4), shaded relief maps (Fig. 4.7), and gray-scale maps (Fig. 4.9).

Humans located in an actual landscape view features differently than when they are shown on a map: the land surface is undulating, vegetation is three-dimensional and has characteristic structures, and objects appear smaller in the distance. Therefore, some GIS are capable of generating perspective views of three-dimensional surfaces, such that the surface is portrayed from an oblique viewpoint. This can be done by means of closely spaced parallel lines (Fig. 4.14a) or a mesh of grid cells (Fig. 4.14b). Aerial photographs or vegetation maps can also be "draped" over perspective views to provide information about land cover.

More recently, the development of virtual reality has provided new tools for landscape visualization from mapped data. Berger *et al.* (1996) used Vistapro, a software package for simulating three-dimensional landscapes from DEMs, to generate animated "fly-bys" of real landscapes. Vistapro renders realistic scenery by level slicing the landscape color palette to realistically shade cliffs and hills. It does this through user-definable variables, a rules system, fractals, and chaotic maths to add non-uniform textures to the rendering. Life-like landscapes are depicted despite a limited number of possible terrain types: water, beach, vegetation, bare, snow, cliff, and buildings. Initial testing and rendering was done with a 7½′ (30 × 30 m cells) DEM of Crater Lake, Oregon (Fig. 4.15).

Fig. 4.15 Landscape scenes with fractal models of terrain type (i.e. trees) rendered from the Crater Lake 7½′ DEM. Courtesy of Paul Meysembourg, Natural Resources Insitute, University of Minnesota, Duluth.

4.4 Topographic applications in ecology

4.4.1 Assessing biotic/topographic relationships

Ecologists have known for some time the importance of topography to plant community distribution, exemplified in monographs by Whittaker (1956, 1960) about the vegetation of the Smoky and Siskiyou Mountains in the eastern and western U.S. Whittaker pioneered the use of gradient analysis, which attempts to describe and understand the distribution of vegetation in response to one or more environmental, resource, and/or temporal gradients. Figure 4.16 shows a gradient analysis in which tree densities are displayed in a two-dimensional abstract space defined by elevation, aspect, and topographic position (Kessell 1979). **Gradient modeling** generates vegetation maps from empirically derived gradient analyses such as those shown in Fig. 4.16 by using known environmental conditions, in this case topographic variables, to predict vegetation occurrence (Kessell 1996). When gradient analyses have been done for each of the plant species occurring in a region, a wide range of derived maps can be produced, such as species density, species diversity, and dominance/

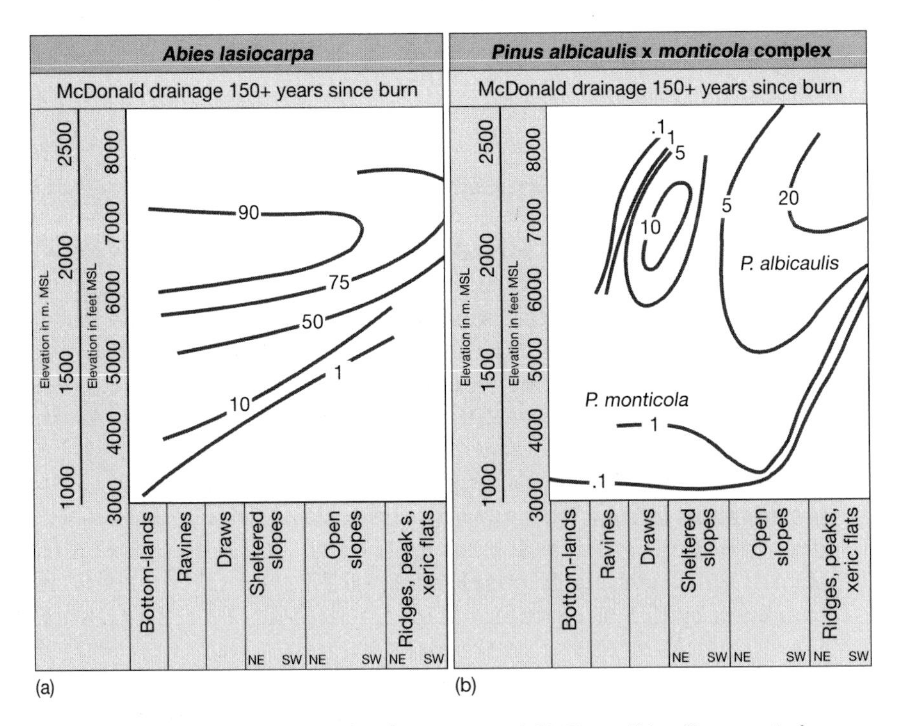

Fig. 4.16 Gradient analyses for: (a) *Abies lasiocarpa*, and (b) *Pinus albicaulis* × *monticola* complex in Glacier National Park relative to elevation, aspect, and topographic position. From Kessell 1996. By permission. *GIS and Environmental Modeling: Progress and Research Issues*. Goodchild *et al.* © 1996, GIS World, Inc.

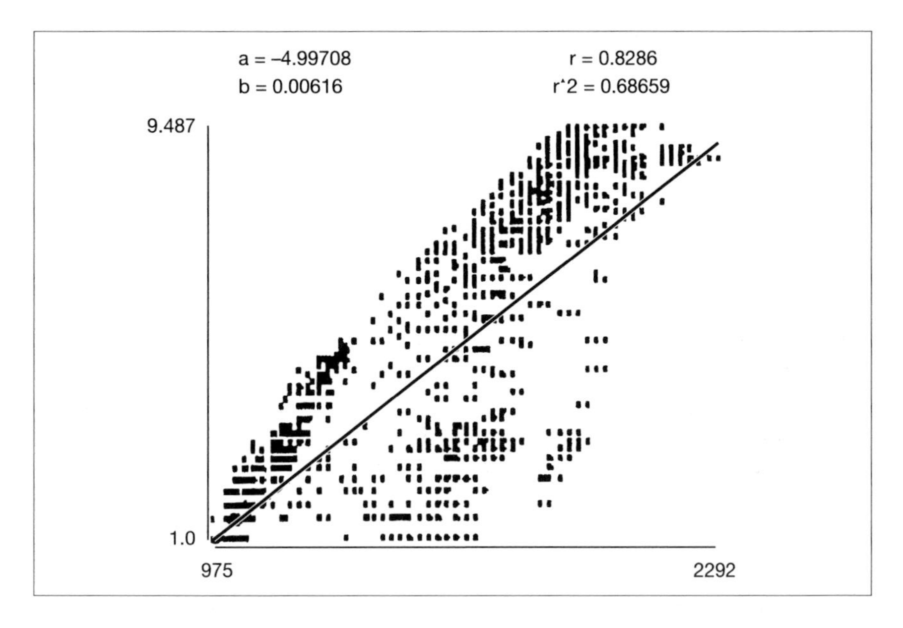

Fig. 4.17 Scatter plot and linear regression of the square root of the relative density of *Abies lasiocarpa* (y axis) versus elevation in meters (x axis), western Glacier National Park. From Kessell 1996. By permission. *GIS and Environmental Modeling: Progress and Research Issues*. Goodchild *et al.* © 1996, GIS World, Inc.

codominance. GIS also can be used to derive new gradient analyses from plant distribution and environmental data layers (Fig. 4.17), a process currently underway in Australia (Kessell 1990).

Ecologists have used several other empirical methods to study the distribution of terrestrial vegetation in relation to topography, particularly in mountainous areas. Statistical correlations are one of the simplest exploratory tools, used by Duguay and Walker (1996) to evaluate relationships between an index of vegetation greenness derived from satellite imagery (NDVI, normalized difference vegetation index — see Chapter 6) and 11 topographic and topoclimatic variables: elevation, slope, relief, downslope convexity, cross-slope convexity, angle of incidence, orogenic precipitation index (OPI), growing degree days (GDD), slope-aspect index, insolation index, and snow probability. For east- and west-facing slopes, NDVI was highly correlated ($r = 0.77$) with OPI, a simple index computed by GIS as elevation divided by distance from the Continental Divide. NDVI was also highly correlated with GDD ($r = -0.75$) and elevation ($r = 0.75$) for west-facing slopes, but was less well correlated with these factors for east-facing slopes ($r = -0.47$ and 0.51, respectively).

Other statistical techniques used to develop empirical models of biotic/topographic relationships include Discriminant Function Analysis (DFA) and Bayesian Statistics, both of which are explored in more detail in Chapter 10. Lowell (1991) used DFA with soils, topographic, and historical vegetation data to develop a spatial model of ecological succession for an area composed of old fields, cedar glades, and oak–hickory forests. Researchers at the Macaulay Land Use Research Institute in Scotland used a Bayesian approach to model habitat distribution for curlew (*Numenius arquata*) and golden plover (*Pluvialis apricaria*) from satellite imagery and DEMs (Aspinall & Veitch 1993; Aspinall 1994).

4.4.2 Topographic controls on insolation

The previous examples illustrate spatial coincidence, but not causality, between topography and biotic distribution. Insolation drives photosythesis, and therefore has a more direct causative influence on vegetation distribution. In a study of maritime chaparral plant communities in California, Davis and Goetz (1990) modeled insolation from topography, and used the results with other environmental variables in a GIS to predict the distribution of different vegetation types. A USGS DEM with 30-m pixels was used to compute slope, aspect, and the angle to the horizon in eight directions for each grid cell (Dozier 1980). Terms for diffuse irradiance and reflected radiation from surrounding terrain were estimated under specified conditions of atmospheric scattering and transmittance and surface albedo. Instantaneous radiation was calculated at hourly intervals for 3 days in each month, and integrated over the entire month to produce maps of monthly solar radiation for the months of December through June. Calculated March radiation was strongly associated with the pattern of natural vegetation (Davis & Goetz 1990).

4.4.3 Topographic controls on disturbance

Disturbance is an important factor in the maintenance of landscape heterogeneity (Turner 1987a). Topography influences disturbance in a variety of ways, from the relatively direct influence of steep slopes on avalanche occurrence (Delcourt & Delcourt 1992) to indirect influences on the susceptibility of vegetation to disturbance by insects and fire.

Baker *et al.* (1991) developed a spatial model for studying the effects of disturbance on the structure of landscapes, designed to simulate repeated disturbances over hundreds of years, simulate the effects of environmental variability or landscape structure in disturbance initiation and spread, and provide information on quantitative changes in landscape structure. The model estimates disturbance probability based on GIS data layers of topography (elevation, slope, and aspect) and vegetation (vegetation type, stand age), and initializes the potential disturbance within a single cell by choosing randomly among cells containing disturbance probabilities exceeding a minimum value specified by the user. The model was written in the SIMSCRIPT II.5 simulation modeling language, but incorporates tools and capabilities of the GRASS GIS and an external statistical analysis package.

Topographic variables are important in wildland fire models, because of topographic influences on microclimate, soil moisture, and vegetation types. Kessel, working initially in Glacier National Park in the western U.S. (Kessell 1979), and later in eastern Australia (Kessell 1990), intergrated a raster GIS with the North American Fire Behavior Model (Rothermel 1972), which uses slope and aspect, as well as vegetation, fuel, and weather data to predict the spread of fire. More recently, Clarke and Olsen (1996) developed a GIS-based model in which fire spread closely resembles "diffusion limited aggregation," an aggregation process that has been used to explain the growth of snowflakes, mineral crystals, and cities. The model extracts input variables from a DEM, a fuel data layer, and a wind table giving probabilities of fire movement at different compass directions and wind magnitudes. The DEM is used to generate hillslope hydrology for each cell within the analysis areas, derived from: (i) soil properties; (ii) insolation, based on slope, aspect, and local horizon conditions at the appropriate latitude; and (iii) relative position in the watershed computed from Band's (1986) automated basin delineation algorithm (see Section 4.2.7.3).

The magnitude of disturbance by flooding is directly related to topography because of the relationship between elevation and flood depth. Depth of flooding has a strong influence on the vegetation of riverine wetlands (Johnston *et al.* 1997), and can be incorporated into models that simulate wetland vegetation change with altered water levels. For example, Pearlstine *et al.* (1985) developed a vegetation simulation model for a lowland hardwood forest along the Santee River in South Carolina, using elevation data to estimate the effects of river diversion on the distribution of habitat types and a GIS to display the modeled results. Elevation differences of centimeters may have a large influence on the magnitude and spatial extent of disturbance by flooding, so GIS applications of this type require vertical elevation data of much greater detail than that typically provided by a DEM (Brimicombe & Barlett 1996).

4.4.4 Watershed modeling

Watershed modeling predicts the location, direction, and magnitude of water flow within a watershed. The topographic analysis capabilities of GIS, especially flow routing routines (see Section 4.2.7), have greatly facilitated the successful integration of watershed models with GIS.

GIS use has simplified derivation of input variables for watershed modeling. Vieux (1991) used a GIS-based TIN to process the terrain of a small watershed for use in a finite element analysis of hydrologic response units. Steube and Johnston (1990) found that GIS input derivation greatly simplified data preparation and handling for runoff estimation using the SCS curve number approach. GIS-derived input variables have been used by several researchers (Harris *et al.* 1991; Kim & Ventura 1993; Levine *et al.* 1993; Smith & Vidmar 1994) to derive variables for non-point source pollution models.

Several off-the-shelf surface water models have been integrated with a GIS by means of a software interface. Different GIS–hydrologic model combinations that have been tested include GRASS–AGNPS (Engel *et al.* 1993), ARC/INFO–AGNPS (Nawrocki *et al.* 1994; Jankowski & Haddock 1996), and GRASS–SWAT (Srinivasan & Arnold 1994), and ARC-INFO-SWAT (Bian *et al.* 1996).

Ecologists are increasingly using watershed models to investigate how ecosystems affect and are affected by the distribution of water and nutrients throughout the landscape. Researchers in western Montana (Running *et al.* 1989; Band *et al.* 1991) developed a spatially distributed hydroecologic model, which combined a watershed model with a forest ecosystem model to simulate cumulative annual evapotranspiration in a mountainous landscape. Krummel *et al.* (1996) used a non-point source pollution model (ANSWERS) to analyze the effect of changing land cover conditions on soil erosion/deposition and surface hydrology in a midwest U.S. township subject to agriculture intensification. Nawrocki *et al.* (1994) also used a non-point source pollution model (AGNPS) to investigate the influence of beaver ponds on surface water quality and quantity. GIS development will facilitate such interdisciplinary efforts, providing a common tool with which to integrate physical, biological, and sociological models.

4.4.5 Topography and climate change

Predicted global climate change has stimulated research into ecosystem responses and feedbacks to atmospheric warming and precipitation change. Such research by necessity has used models to estimate the spatial distribution of climate change, but current general circulation models (GCM) have an effective resolution of the order of about half the width of the contiguous U.S., inadequate to represent such large geographic features as the Rocky Mountains (Pielke *et al.* 1996). However, mountains can significantly influence local climate and weather, making the topographic analysis capabilities of GIS important for fine-tuning estimates of temperature and precipitation change in mountainous areas (Daly & Taylor 1996; Lakhtakia *et al.* 1996; Pielke *et al.* 1996; Running & Thornton 1996). GIS has also been used to adjust for the effects of elevation on temperature and precipitation data collected at weather stations throughout the U.S. (Kittel *et al.* 1996).

GIS topographic capabilities have been equally important in predicting potential impacts of climate change on biota and ecosystems. DEM data have been used in GIS-based predictions of a variety of ecological responses to climate change: inundation of coastal ecosystems by sea level rise (Lee *et al.* 1991), changes in soil moisture and runoff from forested watersheds in Montana (Nemani *et al.* 1993), and changes in chaparral and yellow pine forest distribution in California (Fairbanks *et al.* 1996).

4.4.6 Other topographic applications in ecology

Topographic data are becoming increasingly important in other ecological applications. Limnologists are beginning to use bathymetric data in their work (Schloss & Rubin 1992). Foresters have begun to incorporate topographic influences on soil water content into their forest inventories and ecosystem models (Sieg *et al.* 1987; Band *et al.* 1991; Mackey *et al.* 1996). Plant ecologists have used elevation data to estimate the effects of gravity on the spatial distribution of seed dispersal (Malanson *et al.* 1996). These applications have been facilitated by the development of topographic operations within GIS. As fine-scale DEM data are increasingly made available by governments and orthophoto quad producers, we can expect our knowledge of ecosystem – topographic relationships to grow.

CHAPTER 5

Linear operations

Linear features often have important ecological functions. Streams and rivers convey water from inland to the sea. Coastlines separate terrestrial from aquatic ecosystems. Animal pathways are corridors of organism movement. Ecotones affect fluxes of materials across the landscape. The quantification of these linear features and their attributes often provides new insights into their ecological function.

Some of these ecologically significant linear features have been mapped for centuries. Maritime and river explorers prepared detailed maps of waterways because of their importance as trade routes. Some linear features of ecological significance must still be delineated (see Section 7.1) or inferred from ancillary data (see Section 4.2.7). New technologies, such as animal collars that contain Global Positioning Systems — see Chapter 8), have provided only recently the tools to map other linear features.

Linear measurements and the determination of fractal dimension were discussed in Chapter 3 and the derivation of stream networks from topographic data was discussed in Chapter 4. This chapter describes additional GIS operations for the classification and analysis of linear objects that possess attributes of their own, rather than being merely topological separators between polygons. These linear objects may be fixed (e.g. pipelines) or ephemeral (e.g. animal movements), and real (e.g. streams) or contrived (e.g. political boundaries).

5.1 Linear segmentation

Linear attributes normally are stored in relation to the strings that they describe. In the example in Fig. 5.1, each individual reach segment in a stream drainage network was classified according to predicted water quality (White & Hofschen 1996). However, linear entities may have multiple attributes that do not coincide spatially with digitized strings (Fig. 5.2). When this is the case, new line segments can be defined by performing an "XOR" combination (see Section 3.3.3.2) between a linear data layer and a polygon data layer. This process splits the original lines into new segments defined by their intersections with lines on the polygon layer (Fig. 5.3), which may be assigned attributes from the polygon layer. For example, an XOR combination between a shoreline map and a soil polygon map would yield a shoreline segmented into sections by soil type (Fig. 5.2a).

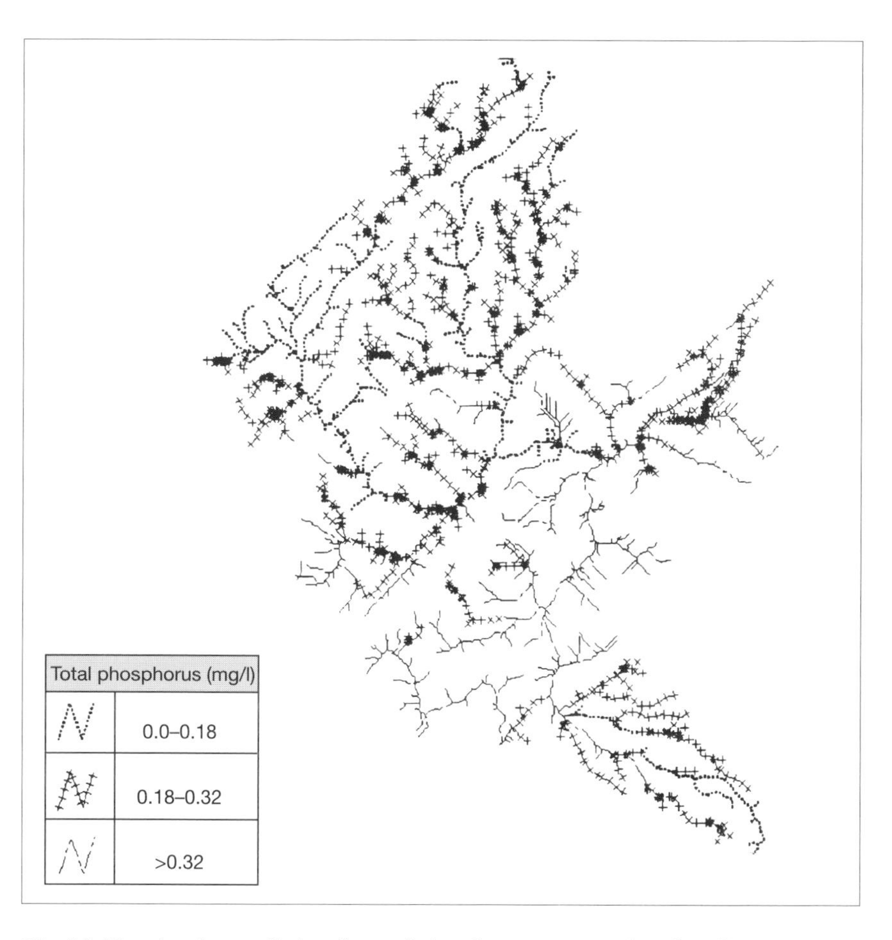

Fig. 5.1 Map showing predictions for total phosphorus concentrations for 1724 reach segments in a drainage basin in northeastern New Jersey. From White & Hofschen 1996. By permission. *GIS and Environmental Modeling: Progress and Research Issues.* Goodchild *et al.* © 1996, GIS World, Inc.

Dynamic segmentation is another line segmentation procedure that divides a linear feature into equal increments, such that each segment can be identified, manipulated, and assigned attributes independently of the others in the overall linear feature. Dynamic segmentation is particularly useful for entities that are straight lines, such as roads and pipelines, which can be represented accurately by relatively few digitized line segments. For example, Maggio *et al.* (1993) used dynamic segmentation to divide a pipeline into 500′ segments, which were classified according to adjacent housing density.

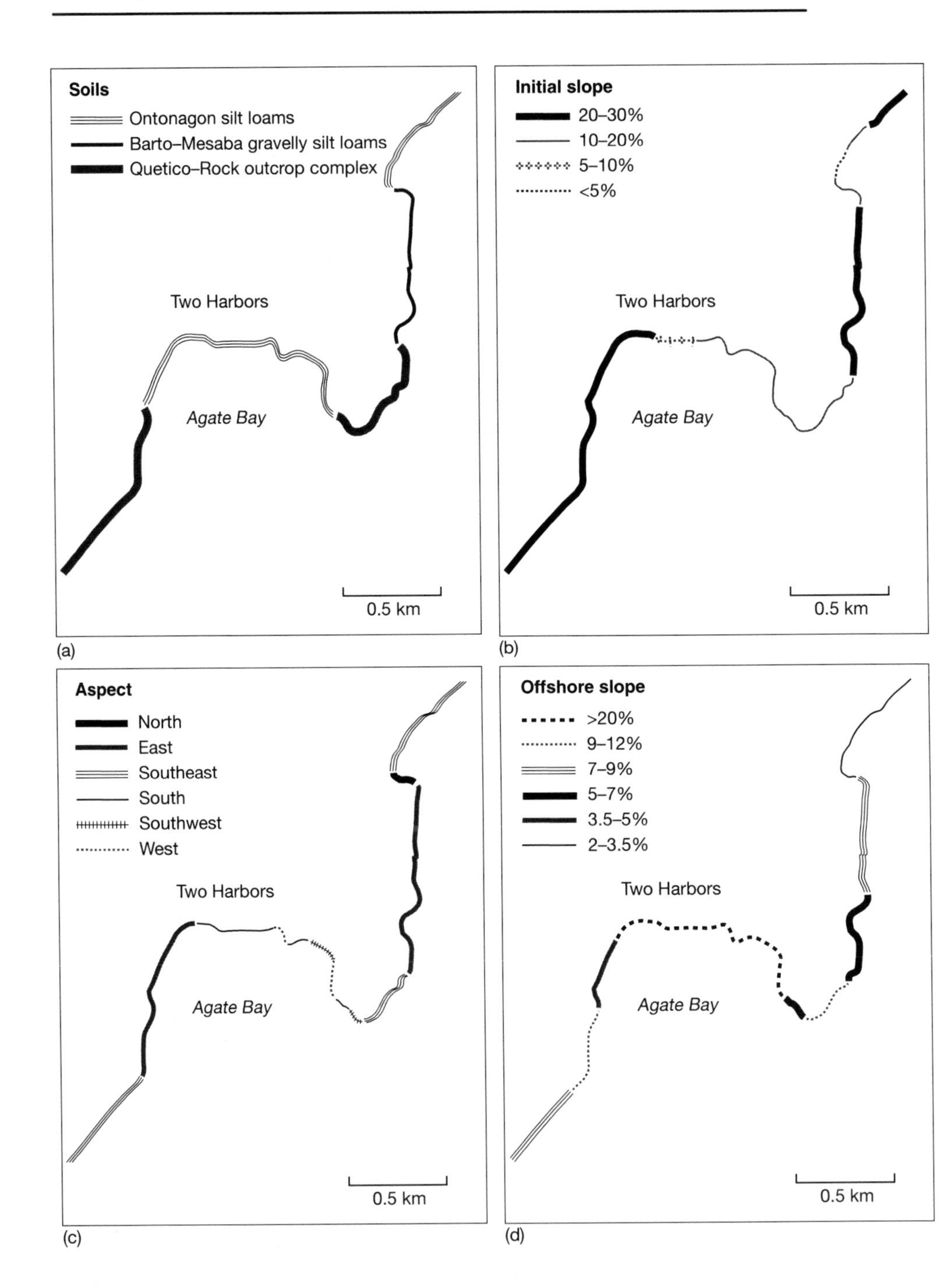

Fig. 5.2 Multiple classifications of Minnesota's Lake Superior shoreline near Two Harbors, Minnesota. Reproduced with permission, the American Society for Photogrammetry and Remote Sensing. Johnston, (1989) Ecological research applications of geographic information systems. In *GIS/LIS '89 Proceedings*, 569–577.

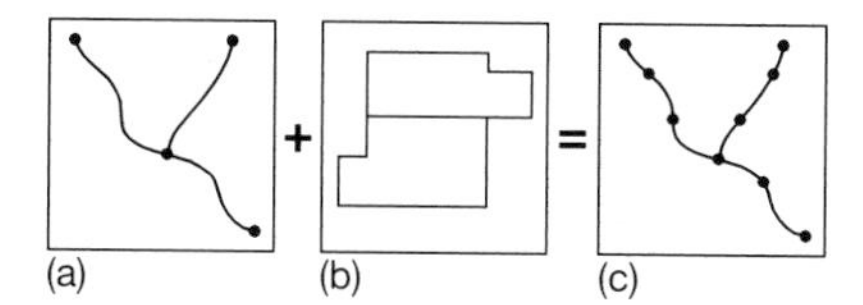

Fig. 5.3 Segmentation of linear features by an XOR combination between linear and polygon data layers. New nodes on the map resulting from the XOR operation (c) occur at the intersection of the linear feature (a) with the polygon boundaries (b).

5.2 Networks

A network is a series of interconnected lines. Networks may be dendritic or rectilinear, as exemplified by stream and road networks. Stream networks branch hierarchically, have unidirectional flow, and are composed of network chains (see Spatial Data Transfer Standard terminology in Table 2.6). Road networks are composed of lines that tend to be straight and intersecting at right angles, have uni- or bidirectional flow, and are composed of complete chains that may or may not form GT-rings (Table 2.6; Fig. 2.9). Different GIS operations are applied to dendritic and rectilinear networks.

5.2.1 Dendritic networks

5.2.1.1 Flow paths

Water flow data layers can be used to determine the path from a designated start location anywhere in the watershed to the drainage network outlet, a capability useful in tracking the path of a pollutant from a point source into and through the drainage network. In a raster drainage network, a start cell is identified and the flow direction data layer (see Section 4.2.7.1) is used to determine the cell-to-cell pathway to the outlet (Jensen & Domingue 1988). To distinguish diffuse from channelized flow, thresholds can be applied to the flow accumulation data layer such that only cells which receive runoff from a sufficiently large drainage area are identified as channels.

A comparable flow path procedure can be done in a vector GIS, except that the topologies of the network chains (i.e. the beginning and ending nodes specified as part of the data structure) are used to determine the directionality of flow from the "start" location to the outlet.

When a pollution source (or other source of material to a stream) is near but not part of the stream network, a **proximity** analysis can be performed in a vector GIS to identify and measure the distance to the closest stream segment within a specified search radius. This stream segment can then be used as the "start" location.

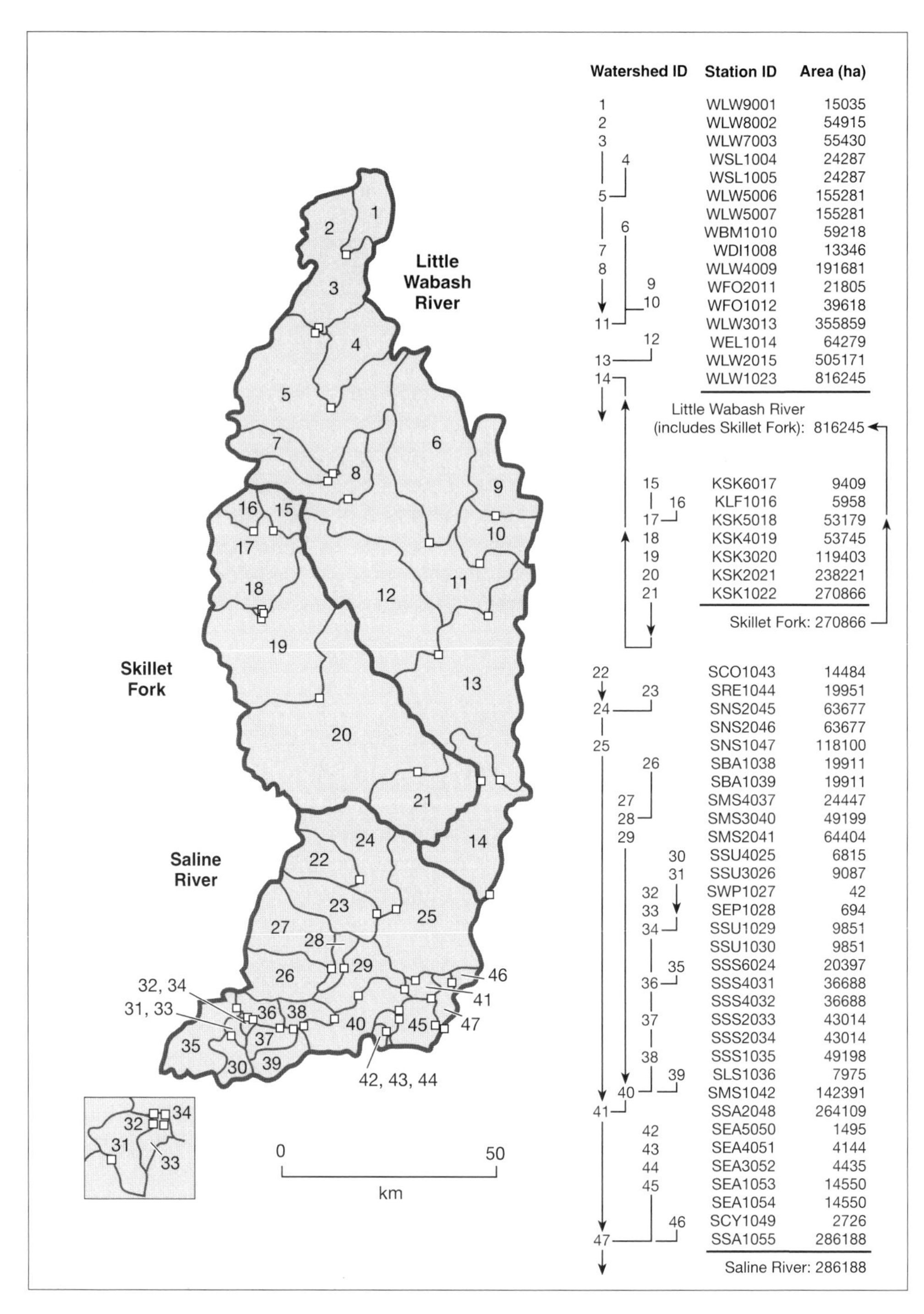

Fig. 5.4 Graph illustrating watershed hierarchy. From Hunsaker & Levine 1995. By permission, *Bioscience*, **45**: 193–203 © 1995, American Institute of Biological Sciences.

5.2.1.2 *Pour point derivation and ordering*

Chapter 4 illustrated the derivation of watersheds from a depressionless DEM (Digital elevation model – see Section 4.2.7.3). These same data layers can be used to determine the pour point where one watershed or subwatershed drains into another. A pour point is the point of lowest elevation on the common boundary between two watersheds, and is determined with a raster GIS as follows (Jensen & Domingue 1988):

1 Generate a data layer of all cells on the border of each watershed using the boundary option of a 3 × 3 moving window (see Section 3.2.5), such that boundary pixels retain the code for their watershed and non-boundary pixels are recoded as zero values. The resultant boundaries will be two pixels wide, showing the watersheds present on both sides.

2 Intersect that data layer with the depressionless DEM and find the minimum elevation of the common boundary for each watershed pair.

In a vector GIS, information on the topology of watershed divides (i.e. the watersheds that adjoin the boundary) is provided by the data structure, eliminating the need for step 1.

The elevations of pour points can be used to determine their hierarchy in the drainage system by ordering them from highest to lowest. This can be displayed graphically as in Fig. 5.4.

5.2.1.3 *Stream ordering*

Stream ordering is a method of assigning a numeric order to chains in a stream network based upon the number and arrangement of tributaries. Stream order has been demonstrated to be related to numerous characteristics and processes of river ecosystems. In stream ordering, headwater streams (those that receive water only from overland flow) are assigned to order number one, and chains in a downstream direction are incremented based on two different numbering methods, described below.

In the Strahler (1957) method, which is used most commonly in ecology, stream order increases by one when two chains of equivalent order intersect (Fig. 5.5a). When two headwater (first order) streams intersect, they form a second order stream. When two second order streams intersect, they form a third order stream, etc. Order number does not increase if a joining stream has a lesser order, only if they are of equal order.

In the Schreve (1966) method, the stream orders are additive (Fig. 5.5b). As with the Strahler method, the intersection of two first order streams results in a second order stream. However, the juncture of a first and second order stream results in a third order stream (1 + 2 = 3), the juncture of a first and third order stream results in a fourth order stream (1 + 3 = 4), and the juncture of a third and fourth order stream results in a seventh order stream (3 + 4 = 7).

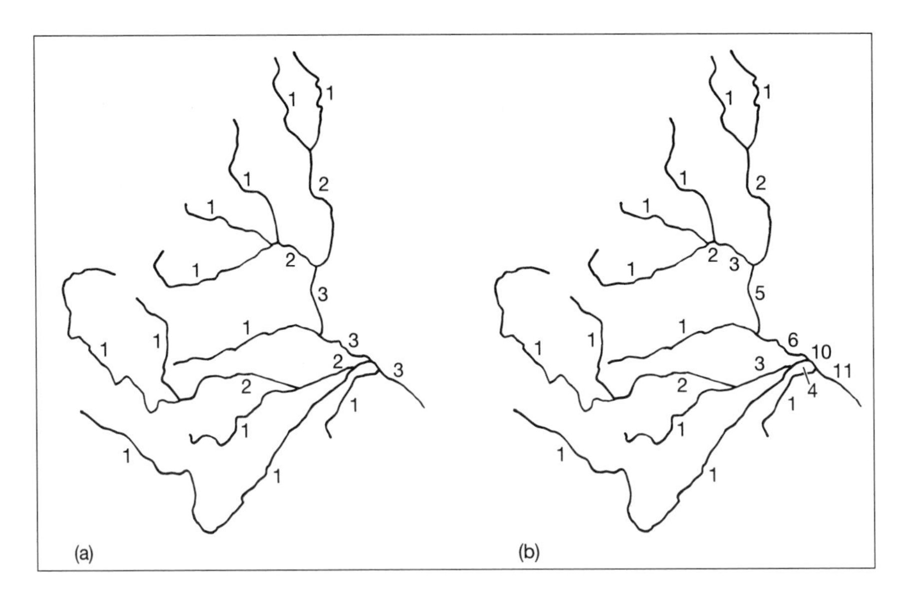

Fig. 5.5 Stream ordering methods applied to the Tittabawassee, Chippewa, and Pine Rivers in Michigan: (a) Strahler (1957) ordering method, (b) Schreve (1966) ordering method.

Because both stream ordering schemes begin with headwater streams, the threshold used to distinguish overland from channelized flow is critically important. Figure 5.6 illustrates stream orders derived from the streams shown on a 7½′ U.S. Geological Survey (USGS) map. Note that third order streams are the maximum attained in this 135-km^2 watershed. Compare this with the DEM-derived drainage network for a 5-km^2 watershed shown in Fig. 4.9. Stream orders obtained from an analysis of this drainage network would attain a much higher order number because no threshold has been applied to the flow accumulation data layer to distinguish cells which have sufficient flow to form a channel. Stream orders obtained from the drainage network for this 5-km^2 watershed would therefore not be comparable to those obtained from the 135-km^2 watershed.

5.2.2 Rectilinear networks

Rectilinear networks are of great importance to human-oriented GIS applications, and a variety of sophisticated analysis functions exist for road transportation, particularly in vector GIS. Rectilinear networks exist in nature in the form of animal trails, but rarely are mapped. Rectilinear networks are composed typically of physical pathways that guide movement, but they can also be the inferred links between grid cells on a raster data surface.

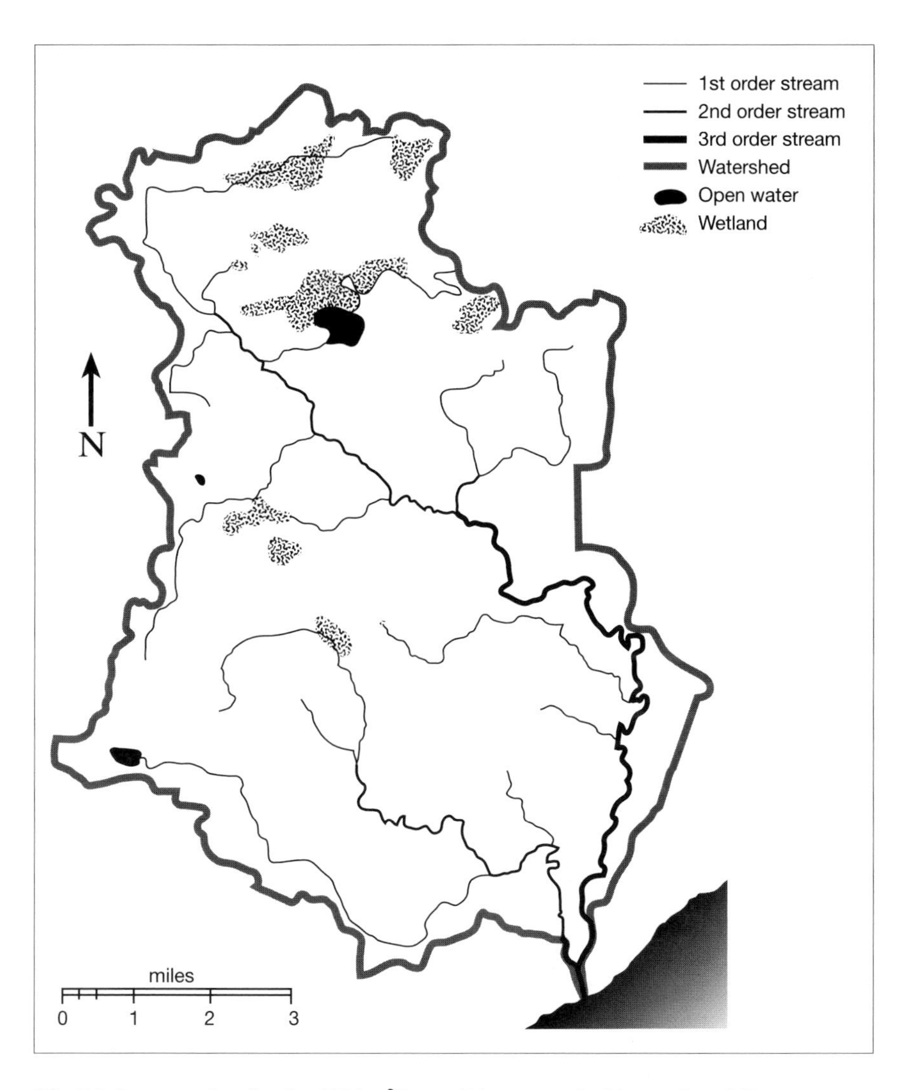

Fig. 5.6 Stream orders for the 135-km^2 Lester River watershed in northern Minnesota. From Johnston *et al.* 1991.

Allocation operations assign road segments to different service centers (Fig. 5.7a), such as the portions of a city which should be serviced by a school based upon student travel time (Lupien *et al.* 1987). Factors that may influence the analysis include impedence (e.g. speed limits, traffic delays), demand (e.g. the number of students on each road segment), barriers (e.g. road blocks), and allowable flow directions (e.g. turns that are permitted at each intersection). An allocation analysis may be uncon-

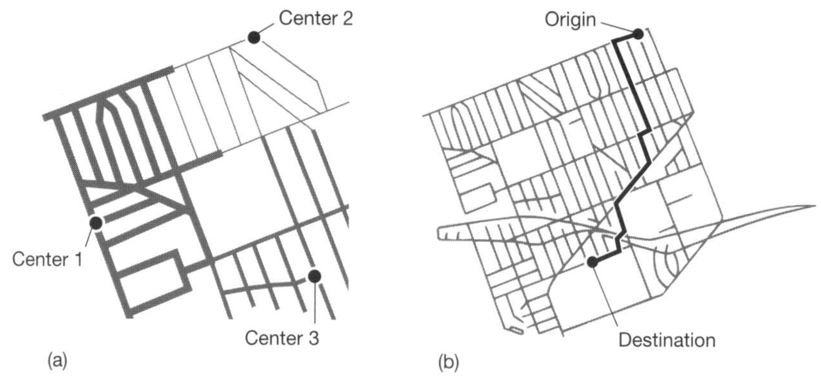

Fig. 5.7 Vector road network operations: (a) allocation of roads to service centers, (b) path-finding procedure to determine the optimal route between an origin and a destination. Reproduced with permission, the American Society for Photogrammetry and Remote Sensing. Lupien *et al.* (1987). Network analysis in geographic information systems. *Photogrammetric Engineering and Remote Sensing*, **53**, 1417–1421.

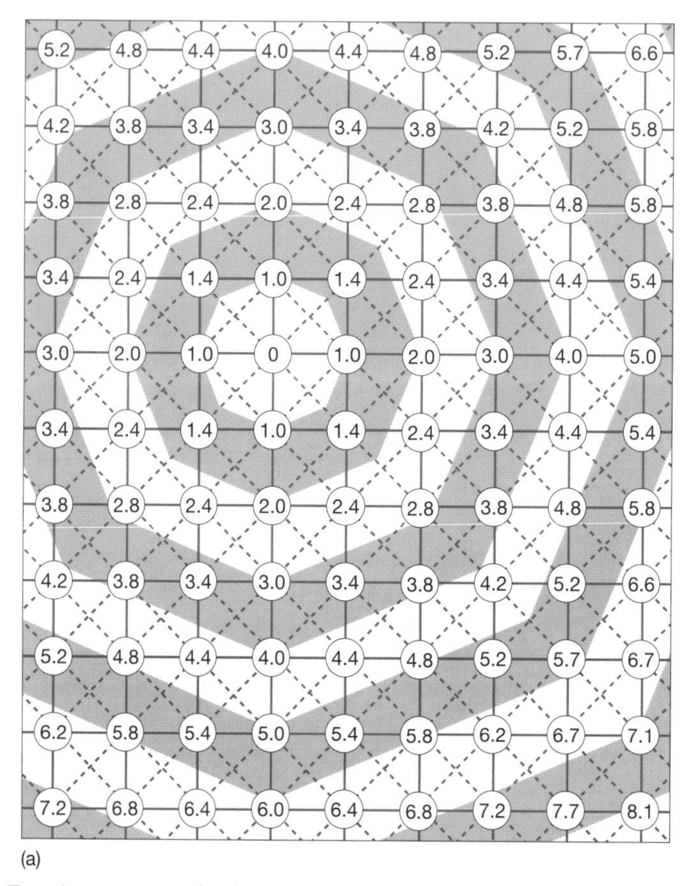

Fig. 5.8 Travel cost computation in a raster GIS: (a) travel costs solely a function of linear distances from a point of origin (0 value), (b) (*opposite*) costs weighted by travel difficulty in different parts of the analysis area. *Geographic Info Systems and Cartographic Modeling* by Tomlin, © 1990. Reprinted by permission of Prentice-Hall, Inc., Upper Saddle Rivers, NJ.

strained, or may be constrained by the capacity of a service center (e.g. the number of students that can be accommodated).

Pathfinding operations determine the optimal path between two or more nodes in a road network, such as the optimal route for a delivery vehicle (Fig. 5.7b). As in the allocation operation, input variables can include impedence, barriers, and allowable flow directions. Stop characteristics (e.g. stop location, duration of stop, whether the stop is for a pick-up or a delivery) may also be considered. In addition to providing the optimal route location based on the input data, some pathfinding programs can also provide street names and turn directions for the route recommended. Travel costs can be assigned to each chain in the road network, based upon distance multiplied by vehicle operation and labor cost per unit length. Expected sales receipts are subtracted from this cost to yield a value representing the true benefit of travel along the route (Lupien *et al.* 1987).

Travel optimization can also be done with a raster GIS, using the eight directions of travel surrounding the focal grid cell in a 3×3 window, such that travel is not constrained to a physical pathway (Tomlin

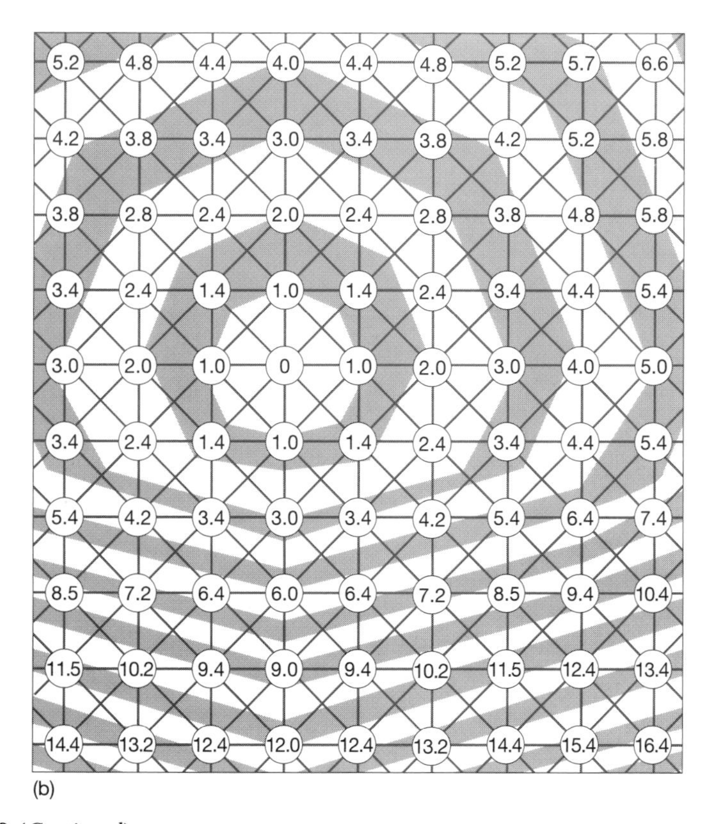

Fig. 5.8 (*Continued*).

1990). In the initial example (Fig. 5.8a), travel costs are solely a function of distance. In the second example, however, travel costs per unit distance are varied in different parts of the analysis area (Fig. 5.8b). This could represent the energy costs of a beaver traveling from a lodge site, the point of origin. Incremental travel costs are set to a value of one where the beaver can swim in the pond surrounding the lodge, but are increased to a value of three where the beaver must travel overland (beavers are much more efficient swimmers than hikers). This is comparable to the idea of impedence values in a vector GIS transportation analysis. Absolute and partial barriers to travel can also be incorporated into the cost analysis (Fig. 3.15). Because animal movements are not restricted to defined pathways, raster GIS methods for travel cost analyses are probably more useful to ecological applications than are the transportation network type of analyses provided by a vector GIS.

CHAPTER 6

Temporal change

The previous chapters have described techniques for analyzing GIS data from a single time period. But many important ecological questions involve the analysis and prediction of change over time: succession, disturbance, animal movements, or material fluxes. This chapter discusses how temporal change can be quantified using a GIS and used to predict future changes with Markov models. Chapter 10 discusses ways in which other types of models can be used to predict temporal change by simulating ecological systems.

Changes are time dependent, therefore time and change are strongly related. There are two approaches to temporal reasoning: change-based and time-based (Al-Taha & Barrera 1990). The change-based approach concentrates on recording changes of facts valid at a certain point in time, without explicitly considering the time domain. The time-based approach considers time as a separate dimension, similar to a one-dimensional space, which uses either time points or intervals as primitives. This chapter focuses on change-based temporal analyses; readers interested in the time-based approach should consult Al-Taha and Barrera (1990).

6.1 Temporal analysis and meterological phenomena

New technologies are generating spatially explicit data about meterological phenomena that influence ecosystem processes. The U.S. National Oceanic and Atmospheric Administration (NOAA) has begun distributing CD-ROMs showing multitemporal snow cover databases derived from satellite imagery and has developed a Doppler radar system capable of quantifying precipitation at a spatial resolution of 4 km^2 for most of the conterminous U.S., both of which will provide important data for hydrologic modeling (Klazura & Imy 1993). In 1985 the U.S. Bureau of Land Management (BLM) began operating a network of radar detectors to triangulate the location of lightning strikes, known as ALDS. Data on lightning are important to ecological researchers because of their relationship to wildfire ignition, particularly in arid regions. Wells and McKinsey (1993) used ALDS data to analyze the average number of lightning strikes per month between 1985 and 1990 in San Diego County, California, and concluded that the period of maximum lightning activity is during the summer prior to the onset of southern California's most severe fire weather. They also demonstrated an increase in the number of lightning strikes with elevation by comparing the location of lightning strikes with a digital elevation model (DEM — see Chapter 4).

6.2 Cumulative multitemporal analyses

Cumulative multitemporal data often provide much more meaningful ecological information than do data from a single observation. In terms of plant growth, for example, annual growing degree-days are more meaningful than the temperature on any single day, and soil moisture surplus/deficit is more meaningful than daily precipitation (Nesbit & Botkin 1993). Continuous data surfaces (i.e. data classified with a ratio scale — see Chapter 2) representing unitemporal observations can be combined into cumulative indices by using a GIS to sum values across individual data layers.

This approach was used by Goward and colleagues (1986) to study seasonal changes in the normalized difference vegetation index (NDVI) for North America, which is related to green vegetation biomass. NDVI values were computed from Advanced Very High Resolution Radiometer satellite imagery (AVHRR — see Chapter 9) as:

$$\text{NDVI } (\text{NIR} - \text{visible}) / (\text{NIR} + \text{visible}) \qquad (6.1)$$

using data from the visible and near-infrared (NIR) channels (1 and 2, respectively) of daily AVHRR imagery. By summing NDVI values from throughout the growing season, Goward *et al.* (1986) were able to accurately predict net annual primary productivity for major North American biomes.

6.3 Multitemporal analyses in image classification

Phenological information provided by multitemporal data can increase the accuracy of image analysis. For example, the North American wetland deciduous tree species *Fraxinus nigra* typically sheds its leaves in autumn several weeks before upland deciduous tree species shed theirs. Wolter *et al.* (1995) took advantage of this phenological characteristic in classifying forest vegetation. They supplemented a summer Landsat Thematic Mapper (TM) image with a mid-September Multi-Spectral Scanner (MSS) image to identify deciduous forest that had lost its leaves by the later image date, thus locating stands of *Fraxinus nigra*. Similarly, they supplemented the TM image with a May MSS image to identify stands of *Populus tremuloides* (a species that leafs out early in the spring), an October MSS image to identify *Quercus rubra* and *Quercus ellipsoidalis* (late deciduous species), and a February MSS image to identify *Larix laricina* (a deciduous conifer).

On a national scale, phenological data are being used by the U.S. Geological Survey (USGS) to generate land cover data for the conterminous U.S. from AVHRR composites (Loveland *et al.* 1991; Eidenshink

1992). Biweekly maximum NDVI values computed from daily AVHRR scenes acquired from March to October 1990 are combined using a multitemporal classification approach to distinguish seasonally distinct spectral-temporal classes. These classes have a unique phenology, including onset, peak, and seasonal duration of greenness. The advantage of this approach is that phenological variations indicated by NDVI trends greatly enhance the ability to distinguish different vegetation types using AVHRR imagery.

A more sophisticated technique for classifying multitemporal imagery involves the use of a spatial time series analysis (TSA) based on Standardized Principal Components, in which multiple images are analyzed as a group (Eastman 1992b). TSA provides both spatial and temporal information about the change events during the time series (Ord 1979). The spatial output of TSA is a series of component images, numbered in order of the amount of variance of the original image set described. To the extent that a pattern is described by a component image, it is then removed from subsequent images. Because of this, subtle change patterns can be detected that would otherwise be obscured by larger, more seasonally related change patterns.

Time series analysis of NDVI data from 36 AVHRR images for the African continent was used to detect vegetation anomalies related to significant El Niño/Southern Oscillation events (Eastman & Fulk 1993). The first component was found to represent the characteristic NDVI integrated over all seasons, whereas components 2–4 were related to seasonal changes in NDVI. The fifth and sixth components were related to the satellite sensor itself, rather than the vegetation. The seventh and eighth components clearly illustrated the 1987 drought in southern Africa caused by El Niño, although this effect was difficult to detect by eye from the original NDVI images or a time profile graph of NDVI values. The technique was shown to be a comprehensive indicator of change events that was sensitive to periodic and aperiodic events alike.

6.4 Change detection

Change detection is the process of comparing spatially explicit databases from two different time periods to determine the location and nature of changes over time. Numerical data, such as summary statistics of forest products logged from a given management area, may reveal that a change has occurred, but change detection with a GIS reveals the location and spatial extent of change.

6.4.1 Change detection techniques

Field ecologists traditionally detect change by statistically comparing data collected at different times and determining whether the magnitude of the change is sufficient to meet a test of significance. Statistical techniques are

Table 6.1 Matrix classification for two data layers. Labels along the side are the class codes for the first data layer, while labels across the top are the class codes for the second data layer. Numbers in the matrix are the codes generated by the analysis.

	1	2	3	4
1	1	2	3	4
2	5	6	7	8
3	9	10	11	12

usually necessary because measurement of an entire population is impossible, requiring a **sample** of the population to be selected. GIS change detection differs from this approach in that a GIS database represents the **entire** population, rather than a sample thereof. Thus, the concept of a statistically significant change is generally not applicable in GIS change detection, because populations are being compared in their entirety. Provided that databases representing an area at different times are perfectly and consistently mapped, **any** change detected between them is significant, regardless of its magnitude.

Change detection is done by determining the **spatial coincidence** of features on sequential maps. The analysis is similar to the Boolean "AND" operator, but generates a data matrix that is not commutative because the data layers are sequentially ordered. For example, the result of the Boolean operation "*A* AND *B*" would be the same regardless of whether *A* is on layer 1 and *B* is on layer 2, or *A* is on layer 2 and *B* is on layer 1, but the same is not true for change detection. The data matrix from a change detection analysis gives the area of land in all possible combinations of the two input data layers (i.e. the product of the number of classes in each).

Table 6.1 lists the possible class outcomes for a matrix analysis between data layers containing three and four classes, respectively. The total number of possible class permutations is 12 (three classes in data layer $A \times$ four classes in data layer *B*). Note that the result of class 1 changing to class 2 (matrix class 2) is different from the result of class 2 changing to class 1 (matrix class 5). This type of analysis is often used to analyze transitions over time, such as vegetation changes caused by disturbance. The information generated by the matrix analysis is useful in itself, or transition probabilities computed from sequential GIS data can be exported to modeling programs designed to predict future changes based on past trends.

6.4.2 Data needs for change detection

The caveat stated above, that databases used in change detection need to be mapped perfectly and consistently, is an important one. Differences between maps, even those that accurately represent the same features, can be due to a variety of causes unrelated to the change in question: differences in projection, georeferencing, data structures, classification, or scale. Map errors compound these differences.

The use of existing databases for change detection is extremely risky, because of the many possible sources of variation between data sources that are unrelated to actual change. For example, the location of swamp conifers mapped at three different times (1865, 1928, and 1956) by three different government agencies is illustrated in Fig. 6.1. Curtis (1959), who used these maps in his book, *Vegetation of Wisconsin*, stated that "slight differences in the location of conifer swamps on the three maps are apparently due to differences in interpretation by the three sets of surveyors and are not believed to represent actual vegetation changes," because the location of conifer swamps is controlled by soil conditions (i.e. saturated organic soils) that would remain relatively stationary over the 100-year time period spanned by the three maps. However, a change detection analysis performed using these three maps would erroneously show large differences in the location of swamp conifers. For example, the swamp conifer stand in the lower right quadrant of Figs 6.1a and 6.1b has a similar shape on both maps, implying that it is the same stand, but its location is completely offset on the two maps (Fig. 6.2). The fact that this wetland does not appear at all on the Forest Inventory map (Fig. 6.1c) may indicate a difference in interpretation, rather than an actual disappearance of the swamp conifers.

Another problem with using existing databases for change detection is that mapping conventions and classifications are often different. For example, at least three different land use maps exist for the U.S. state of Michigan, each prepared from satellite imagery or aerial photography acquired during the 1970s and 1980s. Such data could conceivably be used for analysis of land use change, but the diversity of methods used to create these databases makes change detection impractical (Table 6.2; Hudson & Krogulecki 1987).

Satellite remote sensing is an important data source for temporal analyses using a GIS (Iverson & Risser 1987). A major advantage of remote sensing is its frequent repeat rate, conducive to multitemporal analyses. As in all change detection analyses, however, care must be taken to ensure that differences between dates are not artifacts of the source data. The same classification system and image analysis algorithms should be used for all images. Extreme care should be taken in georefencing and registering scenes from multiple dates, so that "changes" are not artifacts of misregistration. Spectral differences between images unrelated to the sensor target can be caused by degradation of the satellite signal caused by clouds, haze, dust, or other attenuating factors, as well as

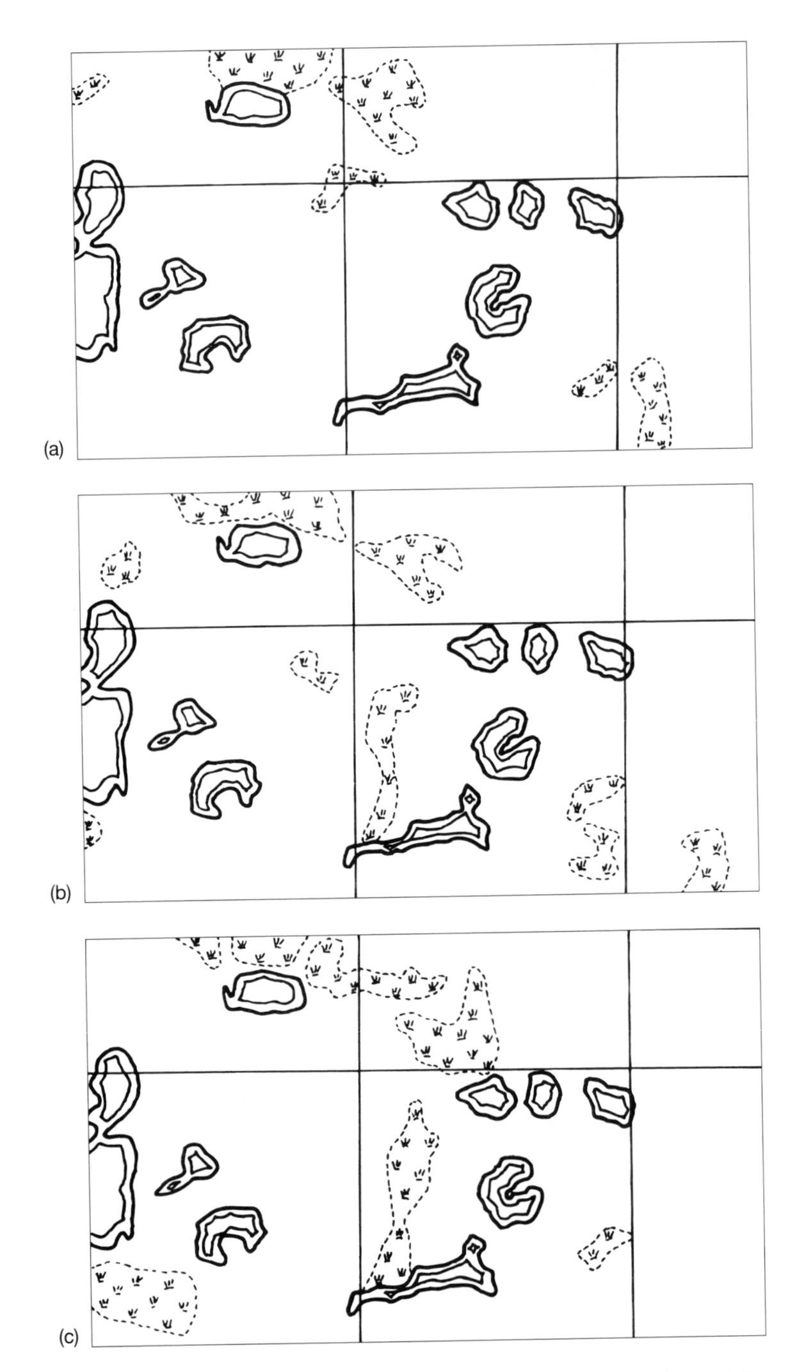

Fig. 6.1 Lakes and wetlands of Plum Lake Township, Wisconsin: (a) Public Land Survey Records, 1865, (b) Wisconsin Land Inventory, 1928, (c) Wisconsin Conservation Department Forest Inventory, 1956. From Johnston 1977, after Curtis 1959.

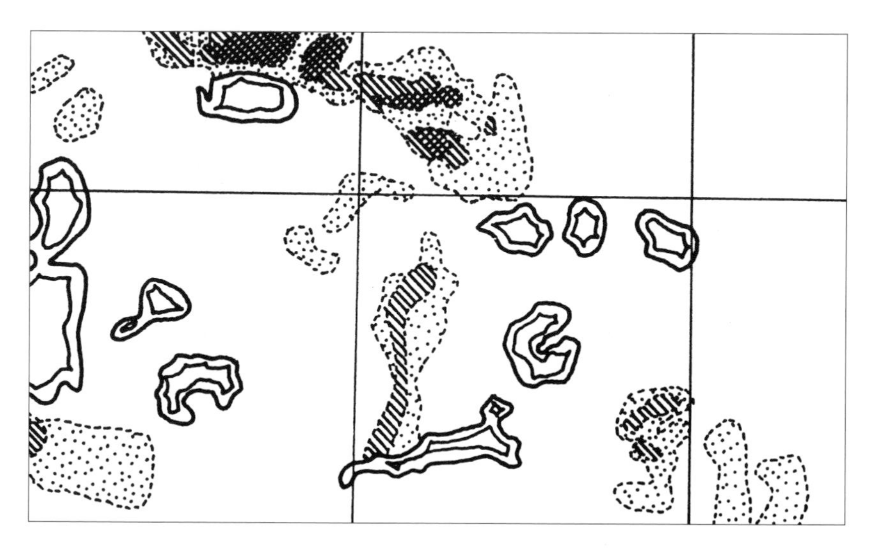

Fig. 6.2 Comparison of wetland maps for Plum Lake Township, Wisconsin. Stippled pattern = wetland appears on only one map. Diagonal-hatching = wetland appears on two maps. Cross-hatching = wetland appears on all three maps. From Johnston 1977.

differences in viewing geometry (solar angle, solar azimuth, and satellite zenith). These differences may be large, as indicated by the study by Eastman and Fulk (1993); sensor-related effects (components 5–6) contributed more to the variance of the original image set than did effects from El Niño (components 7–8). Such sensor-related effects should be eliminated prior to ecological change analysis.

Aerial photography has a less frequent repeat rate than satellite imagery, but provides a longer historical record for temporal analyses because of its availability starting in the late 1930s for most of the U.S. Like satellite imagery, aerial photography is subject to differences in viewing

Table 6.2 Comparison of land use/cover databases for the U.S. state of Michigan. Reproduced with permission, the American Society for Photogrammetry and Remote Sensing. Hudson and Krogulecki, (1987). A comparison of the land cover and use layer of three statewide geographic information systems in Michigan. In: *GIS '87 Proceedings*, 276–287.

	Michigan GIS	USGS land use/ land cover	Michigan resource inventory program
Source imagery	1 : 1 000 000 Landsat TM and MSS transparencies	1 : 58 000 high altitude aerial photos	1 : 24 000 aerial photos
Base map scale	1 : 250 000	1 : 250 000	1 : 24 000
Minimum mapping unit	1 km^2	10–40 acres	2.5 acres
Data structure	Raster	Vector	Vector
Number of categories	9	37	64
Database size (Mb)	0.5	11.1	44.2

geometry. The lower aircraft altitudes used decrease atmospheric effects but increase photogrammetric distortion. Variation in human interpretation of photographic features additionally must be considered. Accurate georeferencing of interpreted data is essential, as well as consistent cartographic projections (see Chapter 2).

6.4.3 Ecological applications of change detection

Johnston and co-workers (1993) used GIS analysis of multitemporal data to study long-term landscape alteration by beaver (*Castor canadensis*). By changing the flow of water in the landscape, beaver impoundments convert terrestrial to aquatic systems, increasing landscape heterogeneity by creating a spatial mosaic of aquatic and semi-aquatic patches in an otherwise forested matrix. Maps of beaver ponds prepared from six dates of historical aerial photography taken between 1940 (when the population was low due to overtrapping) and 1986 (when the population was high) were analyzed with a raster-based GIS, and overlay techniques were used to quantify hydrologic and vegetation changes between map dates. The results indicated that not only did beaver influence a large proportion of landscape area (13%), but that the rate of landscape alteration over the first 20 years of recolonization was much higher than the rate of population increase, indicating that beaver colonies were moving to new habitat sites for reasons other than population growth (Johnston & Naiman 1990a). Once impounded, the majority of sites did not change in water level or vegetation over the decadal periods analyzed (Johnston & Naiman 1990b).

An advantage of using spatially explicit data rather than statistical summaries for change detection is that different types of change can themselves be classified and displayed. Jensen *et al.* (1993) used this approach to produce change maps for coastal wetland habitats between 1982 and 1988 using Landsat Thematic Mapper imagery. All possible permutations of change were generalized into four change categories (Fig. 6.3). New developed/exposed land was associated primarily with human influences, whereas new estuarine unconsolidated bottom was due to differences in tidal height between the two images.

Multitemporal analysis of remotely sensed imagery and aerial photography has been used to detect a variety of ecological changes, illustrated by the following examples. AVHRR has been used to track the interface between desert and arable land in the Sahara (Tucker *et al.* 1985) and the southwestern U.S. (Mohler *et al.* 1986) and the shrinking of the snowpack in the Sierra Nevada (Roller & Colwell 1986). Satellite imagery has also been used for detecting changes caused by disturbances such as fire (Clark *et al.* 1986; Jakubauskas 1990), air pollution (Rock *et al.* 1986), and land clearing (Tucker *et al.* 1984; Hall *et al.* 1991). Aerial photography has been used to monitor vegetation succession (Remillard *et al.* 1987; Green

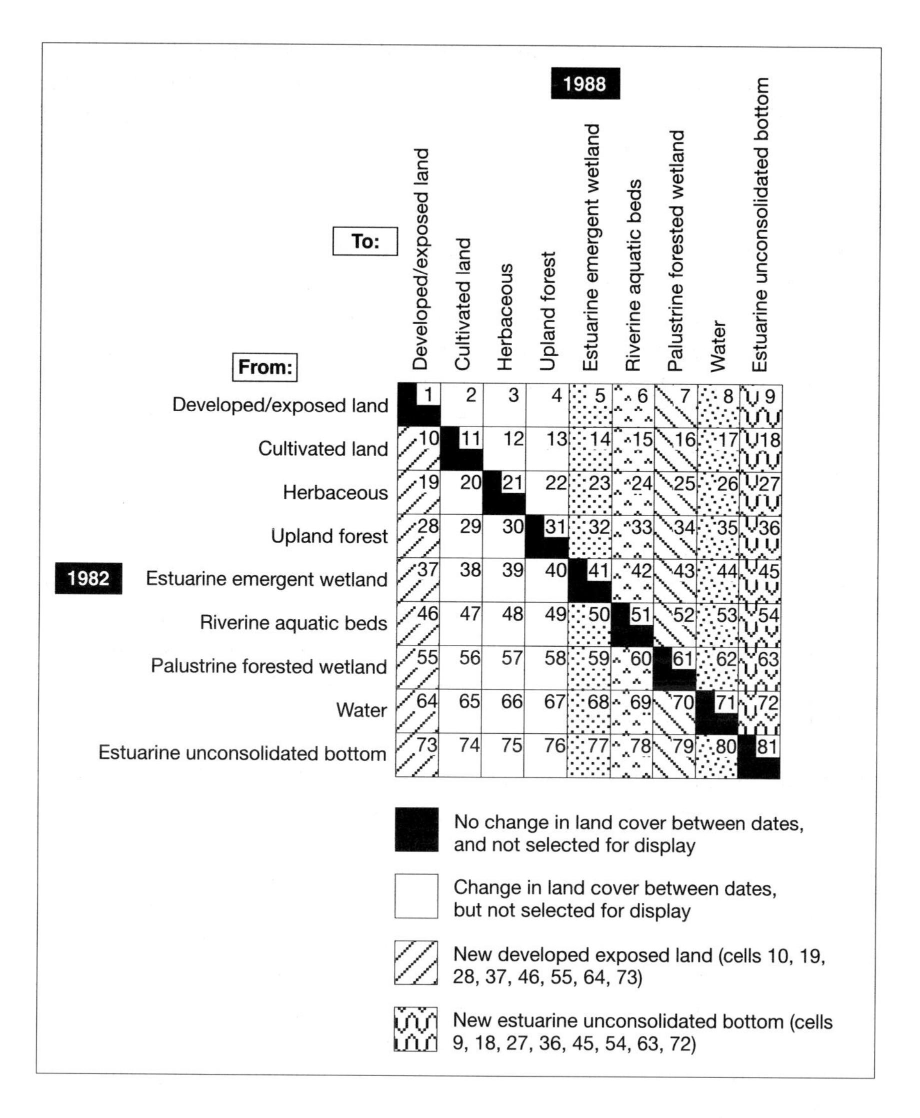

Fig. 6.3 Change detection matrix classifying "from-to" classes for display in a change detection map. Diagonal cells represent areas in which no change occurred. All possible changes are shown, even though some (e.g. water to upland forest) are highly unlikely within the time frame analyzed. Reproduced with permission, the American Society for Photogrammetry and Remote Sensing. Jensen *et al.* (1993). An evaluation of the Coast Watch change detection protocol in South Carolina. *Photogrammetric Engineering and Remote Sensing*, **59**, 1039–1046.

et al. 1993), wetland loss (Ulliman 1992), changes in the distribution of submerged aquatic vegetation (Welch *et al.* 1988; Lo & Ries 1992; Ferguson *et al.* 1993), and changes in land use (Turner 1987b; Ilbery & Evans 1989; Dunn *et al.* 1991).

6.4.4 Computing transition probabilities from change detection analyses

Changes from one land class to another can be described mathematically as probabilities that a given pixel will remain in the same state or be converted to another state. The expected change in landscape properties can be summarized by a series of transition probabilities from one state to another over a specified unit of time:

$$p_{i,j,\tau} = n_{i,j} / \sum_{j=1}^{m} n_{i,j} \tag{6.2}$$

where $p_{i,j,\tau}$ is the probability that a given pixel, or unit of the landscape, has changed from class i to class j during time interval τ, and $n_{i,j}$ is the number of such transitions across all pixels in a landscape of m classes (Anderson & Goodman 1957).

6.5 Markov modeling

When assembled in a matrix and used to generate a temporal series, known as a Markov chain, transition probabilities form a simulation model of changes over time (Pastor *et al.* 1993; Acevedo *et al.* 1996; Johnston *et al.* 1996). Markov models have been assembled to simulate succession of forests (Waggoner & Stephens 1970; Shugart *et al.* 1973; Johnson & Sharpe 1976; Hall *et al.* 1991), heathland (Jeffers 1988), and wetland vegetation associated with beaver ponds (Pastor *et al.* 1993). Early Markov models were parameterized using succession rates measured or observed in the field, but the development of remote sensing and GIS techniques has enabled researchers to calculate less biased transition probabilities from the full extent of the landscape (Hall *et al.* 1991; Pastor & Johnston 1992).

The eigenvalues and eigenvectors of a Markov matrix are useful mathematical properties from which ecological processes can be inferred. They satisfy the equation:

$$A\mu = \lambda\mu \tag{6.3}$$

where A is the matrix of transition probabilities and λ is a particular eigenvalue associated with the eigenvector μ. Because all rows in a Markov matrix sum to 1, the dominant eigenvalue is therefore equal to 1. The dominant eigenvector of a Markov matrix is the steady-state distribution of areas among the various states, and the rate of approach to steady state (proportional to the ratio of the dominant eigenvalue λ_1 to the absolute value of the second largest eigenvalue λ_2) can be used to predict the time required for convergence to steady state (Caswell 1989).

Pastor and co-workers (1993) used Markov modeling to evaluate

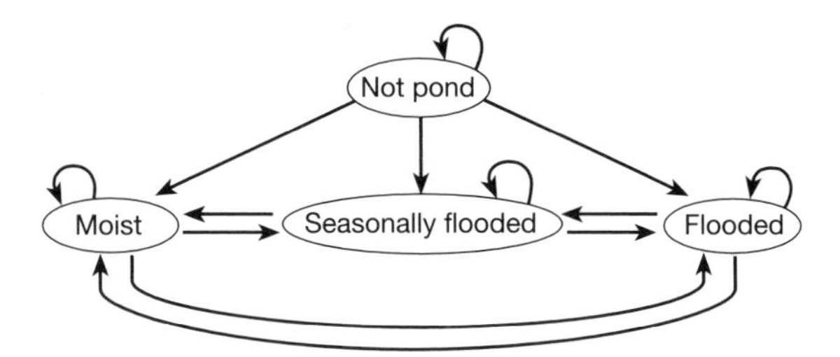

Fig. 6.4 Pathways of land transitions in valley bottoms. Arrows returning to a given class indicate areas which did not change over the time period investigated.

historical landscape changes caused by the beaver pond building described above in Section 6.4.3. Aerial photographs were used to map and classify floodable areas, all land that was impounded by beavers at any time during the 50-year period, into four classes, of which the first three describe soil moisture conditions within beaver ponds: (i) moist; (ii) seasonally flooded; (iii) flooded; and (iv) not ponded. A GIS was used to compute transition probabilities among these four moisture classes (Fig. 6.4). Markov modeling using the computed transition probabilities revealed that although the steady-state distribution among the different hydrologic classes differed by decade, the time required for 95% convergence of the landscape to steady state fell within a narrow range, 30–300 years, regardless of spatial scale, watershed, antecedent soil type (upland, mineral soil wetland, or peatland), or cohort of pond creation (Pastor *et al.* 1993). Plotting the stable acreage fraction of flooded land versus moist meadow (impoundment types 2 and 4 above) for each of the four decades analyzed revealed that the moist class rose and fell in inverse relation to that of the flooded class, and that the variation in acreage among the two classes was almost perfectly correlated ($r = -0.912$). Thus, the extremely complicated patterns seen from the maps were reduced by GIS and algebraic analysis to a simple description of a landscape whose principal dynamic is an alternation of flooded wetlands with moist meadows as ponds are occupied, abandoned, and reoccupied by beaver (Pastor & Johnston 1992).

CHAPTER 7

Spatial interpolation

Sampling techniques provide a subset of information about a universe, providing data for some locations, but not others. Spatial interpolation techniques are designed to fill in the gaps between data points. Those gaps usually constitute the majority of the area mapped, so the spatial interpolation technique used can be critical to the quality of the data.

7.1 Interpolation by boundary determination

Drawing boundaries around areal units that have some common property is a time-honored means of spatial interpolation. The success of this method depends largely on the discreteness of an ecological entity and its boundaries (see Chapter 2). Two steps are involved in this interpolation method:

Step 1 — an area is examined, and classified or sampled;
Step 2 — a boundary is placed to distinguish the area from dissimilar areas.

These steps may be accomplished qualitatively or quantitatively in a variety of different ways.

7.1.1 Human interpolation

Human interpretation is the most common method of spatial interpolation in aerial photograph analysis (see Chapter 9) and soil mapping. A region is examined and areas deemed to meet classification criteria are circumscribed with a polygon boundary. Selection and elimination keys may be used to aid the interpretation by formalizing the thought process (Table 7.1), but the method is basically intuitive.

Figure 7.1 illustrates land use boundaries delineated by eight different aerial photograph interpreters (Gong & Chen 1992). None of the eight is considered to be the true rendition; rather, the overlay of these maps defines boundary zones indicative of the true boundary location. Some of the variation may be due to error, but some may be merely differences in interpretation, which can be substantial if interpreters are not following the same set of guidelines (Bie & Beckett 1973). Interpreter experience, consistent training, and quality checks help to standardize this methodology and reduce error.

Human interpretion is more subjective than other spatial interpolation techniques, and may be perceived as inferior to quantitative

Table 7.1 Air photograph interpretation key for the identification of hardwoods in summer. Reproduced with permission, the American Society for Photogrammetry and Remote Sensing. Sayn-Wittgenstein (1961). Recognition of tree species on air photographs by crown characteristics, *Photogrammatric Engineering*, **27**, 792–809.

1. Crowns compact, dense, large	
2. Crowns very symmetrical and very smooth, oblong, or oval; trees form small portion of stand	Basswood
2. Crowns irregularly rounded (sometimes symmetrical), billowy, or tufted	
3. Surface of crown not smooth but billowy	Oak
3. Crowns rounded, sometimes symmetrical, smooth surfaced	Sugar maple,* beech*
3. Crowns irregularly rounded or tufted	Yellow birch*
1. Crowns small or, if large, open or multiple	
6. Crowns small or, if large, open and irregular, revealing light-colored trunk	
7. Trunk chalk white, often forked; trees tend to grow in clumps	White birch
7. Trunk light, but not white; undivided trunk reaching high into crown, generally not in clumps	Aspen
6. Crown medium sized or large; trunk dark	
8. Crown tufted or narrow and pointed	
9. Trunk often divided, crown tufted	Red maple
9. Undivided trunk, crown narrow	Balsam poplar
8. Crowns flat topped or rounded	
10. Crowns medium sized, rounded; undivided trunk; branches ascending	Ash
10. Crowns large, wide; trunk divided into big spreading branches	
11. Top of crown appears pitted	Elm
11. Top of crown closed	Silver maple

* A local tone-key showing levels 4 and 5 is usually necessary to distinguish these species.

methods. This is not necessarily the case; a well-trained field mapper or aerial photograph interpreter can process complex information about an ecological entity, using a variety of clues to classify an area and distinguish its boundary. Ecological correlates, such as the relationship between landscape position and vegetation or soil type, may be incorporated intuitively into the classification and delineation process. Such reasoning is often difficult to duplicate with more quantitative methods.

For some purposes, human interpolation is considered superior to more quantitative methods. In the U.S., wetland delineation by field inspection is the standard methodology (National Academy of Science 1995). Quantitative techniques such as Prevalence Index (U.S. Army Corps of Engineers 1987) may be used in classifying the contents of a wetland (interpolation step 1 – see above), but more qualitative techniques are employed to locate wetland boundaries (interpolation step 2 – see above).

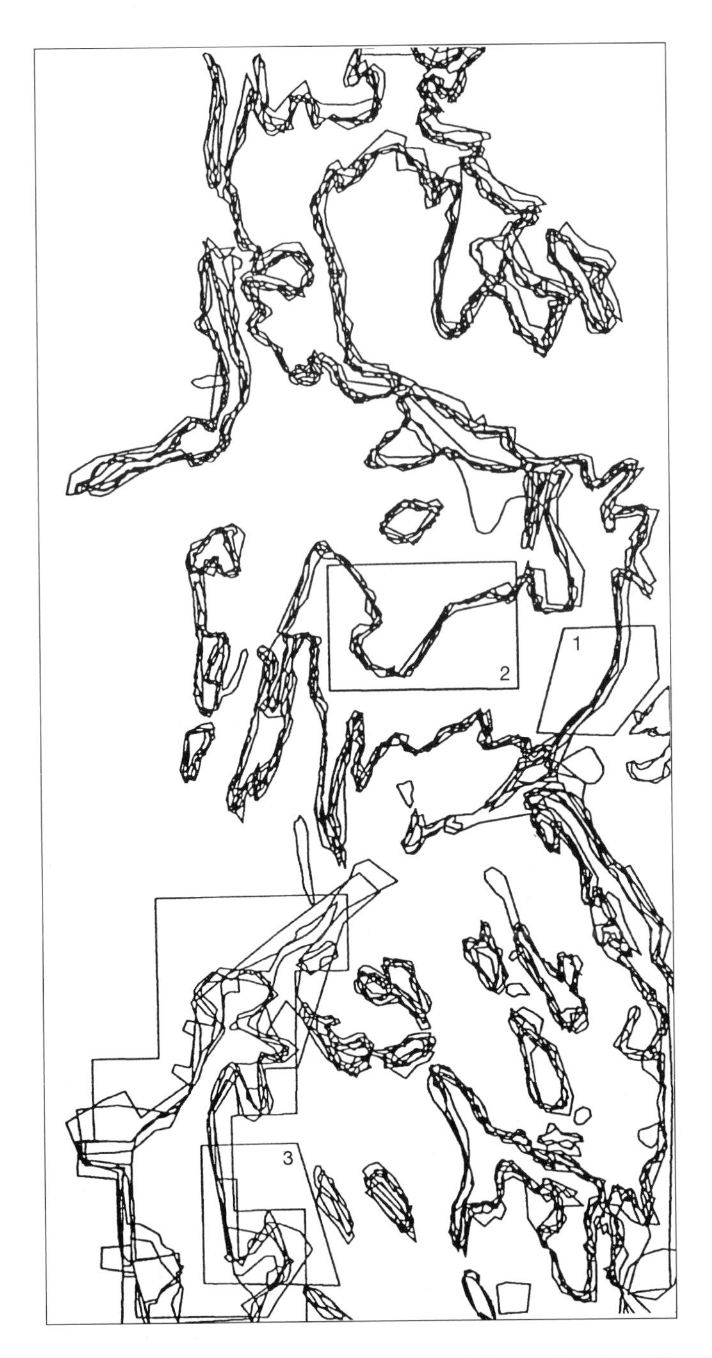

Fig. 7.1 Compilation of boundaries from land use maps delineated by eight different aerial photograph interpreters. Note that there is much higher agreement among boundaries in regions 1 and 2 than in region 3. Reproduced with permission, the American Society for Photogrammetry and Remote Sensing. Gong and Chen (1992). Boundary uncertainity in digitized maps I: Some possible determination methods. In: *Proceedings, GIS/LIS '92*, 274–281.

7.1.2 Quantitative methods for boundary determination

7.1.2.1 Analysis of transect data

Transects are linear arrays of sample points. Transects usually are oriented perpendicular to perceived boundaries, so as to maximize the ecological gradient while minimizing the distance traversed. To determine boundary location, the data obtained from a transect are analyzed for discontinuities in ecological properties.

When suitably located and sampled, transects have the advantage of providing maximum information with minimal effort. Analysis of transect data can provide information on boundary location, width, and contrast. Transect data may be collected at a variety of time and space scales. Most transect data are collected using conventional field techniques, but new technologies are enabling ecological researchers to collect transect data using ground-based and airborne remote sensors (Shih & Doolittle 1984; Gosz *et al.* 1988; Matson & Harriss 1988). Transect data can be aggregated progressively to analyze boundaries at different hierarchical levels (Forman & Godron 1986).

Transect placement and sample spacing intervals are often determined non-randomly, which may bias the results toward boundaries that are readily observable. Also, transect data are one-dimensional; results from transects must themselves be interpolated to create two-dimensional polygon boundaries.

Once transect data have been collected, quantitative techniques must be used to locate and characterize ecological discontinuities (Ludwig & Cornelius 1987). A simple but powerful method for boundary determination is the moving split-window (Webster & Wong 1969). A double window is laid over equal numbers of equally spaced sample points and the dissimilarity (distance) between attribute values in each window half is compared statistically (Fig. 7.2). The window is moved sequentially along the transect until statistical comparisons are computed for the entire length. Boundary locations occur at maximum values of the difference metric, which indicate that the rate of attribute change is at a maximum. Whittaker (1960) was one of the first ecologists to use this approach,

Fig. 7.2 The moving split-window for analysis of one-dimensional data using an eight sample window width (● = sample point). The mean for each window half is computed from the sample values, and the dissimilarity between halves is calculated. The window is then moved along the transect by one sample station, and the process repeated until the entire transect is covered.

calculating similarity coefficients between adjacent stations along a moisture gradient to analyze vegetation in the Siskiyou Mountains.

Various metrics may be used to determine differences between adjacent window halves. The Squared Euclidean Distance (SED) has tended to agree with field observations, and thus is used most commonly (Wierenga *et al.* 1987). The mean attribute value is calculated for each window half (*A,B*), and SED is computed as the square of the difference between the means of each variable in adjacent windows, summed across all variables measured:

$$SED_{nw} = \sum_{i=1}^{a} (\bar{X}_{iAw} - \bar{X}_{iBw})^2 \tag{7.1}$$

where n is a station or mid-point between window halves, w is window width, and a is the number of variables sampled at each station (Brunt & Conley 1990; Turner *et al.* 1991). The window is then moved up the transect 1 station, and SED_{n+1} is obtained, producing a series of values that represent successive differences between window halves along the transect. Other dissimilarity metrics which have been used with the moving split-window method include Student's *t*-test (Webster & Wong 1969), coefficient of dissimilarity (Beals 1969), discriminant functions and Mahalanobis distances (Webster 1973), Hotelling-Lawley trace *F* values (Wierenga *et al.* 1987), and Wilk's lambda (Nwadialo & Hole 1988).

In addition to being a useful tool for boundary determination, the moving split-window technique can also quantify boundary width and strength. When the statistical values (e.g. SED) are plotted against transect position, abrupt boundaries appear as high, narrow peaks, whereas gradual boundaries appear as wider and lower peaks (Fig. 7.3).

Wierenga *et al.* (1987) used the moving split-window technique to distinguish vegetation zones along a 2700-m transect in the northern Chihuahuan desert. Window sizes of six and ten sample points provided similar results, locating six discontinuities defining seven vegetation zones. Although similar peaks appeared using a window width of 2, the greater sample-to-sample noise made the results more difficult to interpret. These vegetation discontinuities were strongly coincident with soil discontinuities determined using the same technique, suggesting that the boundary was edaphically controlled. The moving split-window technique can also be used to identify biogeochemical boundaries that are not visibly obvious (Johnston *et al.* 1992).

7.1.2.2 Textural analysis

Remotely sensed images provide information about the entire landscape, including boundaries. Information about boundaries can be extracted

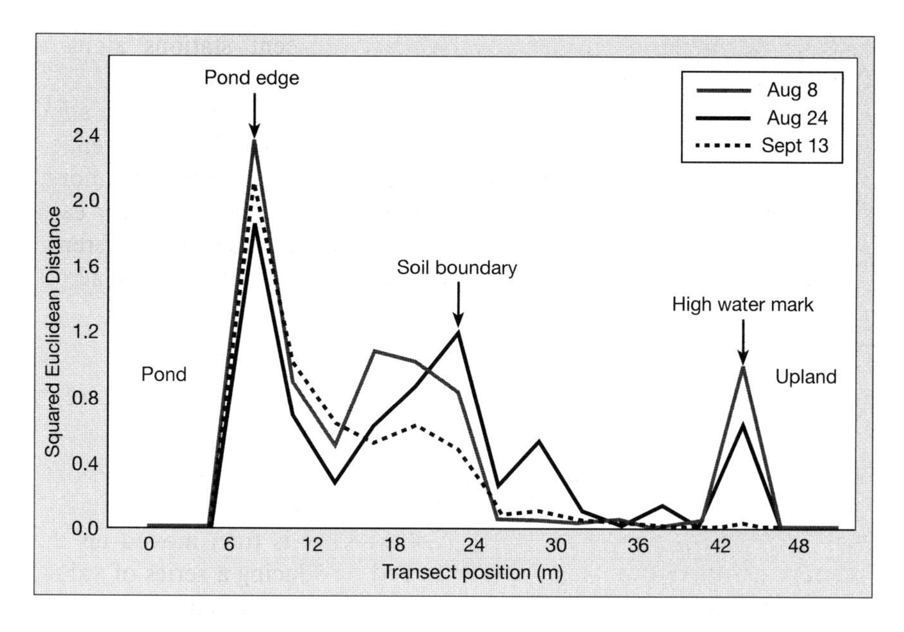

Fig. 7.3 Ecotones determined by moving split-window analysis (window width = 6) of principal components data related to soil water anaerobiosis at 25 cm depth. From Johnston *et al.* 1992.

from the image as a whole by the use of the moving window technique, which scans the image with a two-dimensional window (see Section 3.2.4). The moving window technique can be applied to any two-dimensional digital data, including aerial photography scanned with a video digitizer or scanning camera. Johnston and Bonde (1989) used textural analysis, a moving window technique that measures boundary contrast as relative difference between the reflectance values of picture elements (Musick & Grover 1991), to analyze boundaries within a 324-km^2 Landsat Thematic Mapper image from northern Minnesota (Fig. 7.4). They applied this technique using a map of normalized difference vegetation index (NDVI), a measure of vegetation greenness indicative of biomass production, derived from Landsat Thematic Mapper imagery (see Section 6.2). Boundaries between areas of high and low NDVI appeared as bright borders on the analyzed image, while water and homogeneous vegetation were uniformly dark due to the lack of variation among NDVI values (Fig. 7.4). Nellis and Briggs (1989) performed a similar analysis on the Konza prairie in Kansas. Texture analysis can detect boundaries of varying widths and contrasts because it employs a continuous data surface rather than categorical data. However, the only information textural analysis provides about polygon contents is that there are no large differences in data values within the boundaries identified.

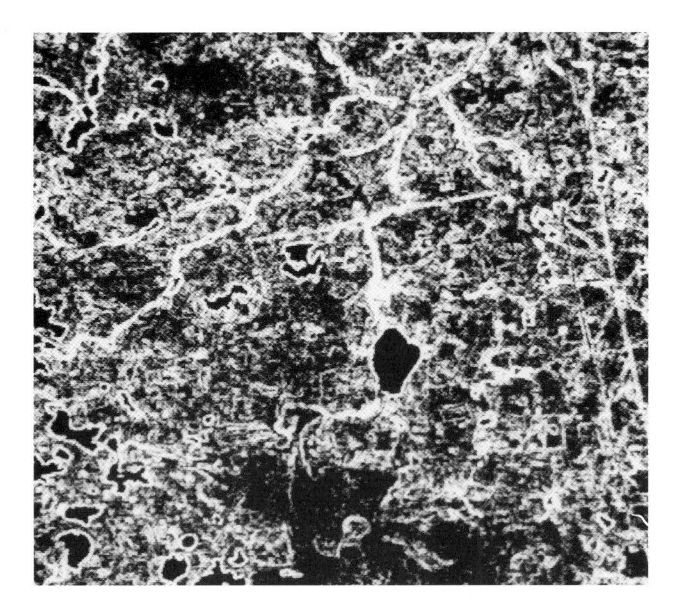

Fig. 7.4 Boundaries between areas of high and low vegetation index, determined by textural analysis of normalized difference vegetation index (NDVI) values derived from a Landsat Thematic Mapper image of north central Minnesota. High-intensity pixels occur where the contrast among vegetation index values in the 3 × 3 pixel scan window was greatest. Reproduced with permission, the American Society for Photogrammetry and Remote Sensing. Johnston, and Bonde, (1989), Quantitative analysis of ecotones using a geographic information system. *Photogrammetric Engineering and Remote Sensing*, **55** 1643–1647.

7.1.2.3 Thiessen polygons

Thiessen polygons (also known as Voronoi polygons or Dirichlet cells) define individual areas of influence around each point in such a way that the polygon boundaries are equidistant from neighboring points, and each location within a polygon is closer to its contained point than to any other point (Maggio & Long 1991). Thiessen polygons can be generated from point data using a three-step process (Green & Sibson 1978; Ripley 1981). The points are joined to their nearest neighbors by lines creating triangles (Fig. 7.5b). Next, the perpendicular bisector of each line is drawn to its intersection with two other perpendicular bisectors, which becomes a node of the Thiessen polygon (Fig. 7.5c). Finally, the initial lines between points and their nearest neighbors are removed (Fig. 7.5d). The division of a region into Thiessen polygons is determined completely by the location of the sample points. Therefore, sample points that lie on a regular grid will define Thiessen polygons of uniform size and shape, whereas sample points that are spaced irregularly will have irregular Thiessen polygons.

There are three disadvantages of using Thiessen polygons for spatial interpolation (Burrough 1986). First, the size and shape of the areas

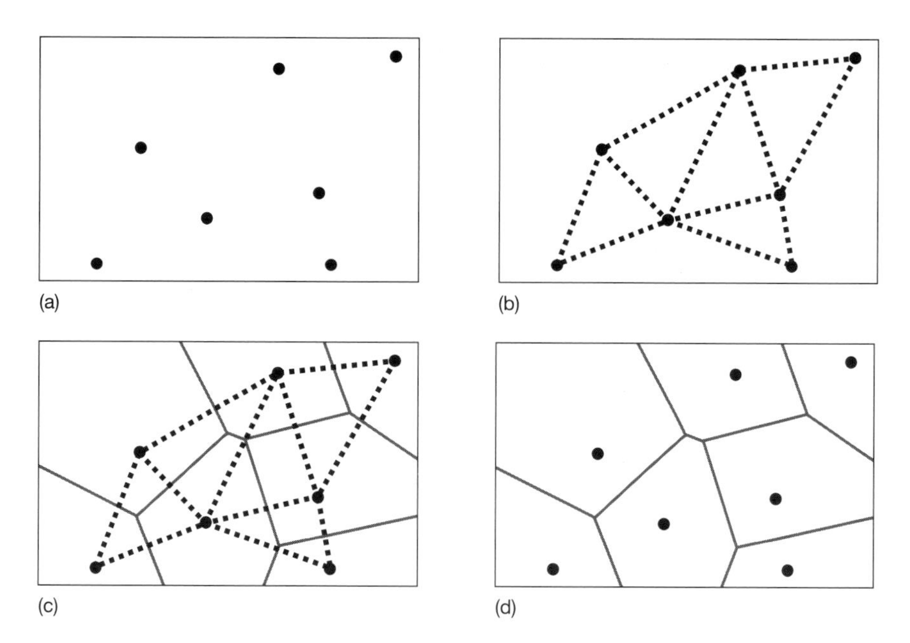

Fig. 7.5 Steps in producing Thiessen polygons from sample points: (a) point locations, (b) points joined by lines forming triangles, (c) perpendicular bisectors of the initial lines are added, (d) initial lines are removed.

depend on sample layout, which can lead to irregularly shaped polygons, particularly at the edges of the area mapped. Second, the value assigned to each cell is based on one and only one value, therefore estimation error cannot be determined. Finally, computation of a value at an unsampled point is a function solely of the polygon within which it is contained, rather than the values of the several points that lie closest to it. The advantage of using Thiessen polygons is that, unlike the other methods described below, the method can be used with categorical (nominal) as well as continuous data (Fig. 7.6).

7.1.2.4 Fuzzy sets

Fuzzy set theory was developed to provide an appropriate way to represent intrinsically imprecise objects in qualitative terms within a mathematical framework (Zadeh 1965). In fuzzy set theory, elements of the universe have grades of membership in a set. This means that elements may be partial members of a set, rather than necessarily being members or non-members. In fuzzy set theory, elements have membership values in the interval [0,1] rather than in the set {0,1} as in classical set theory (Roberts 1989). Fuzzy set theory is a scientifically more realistic way of categorizing features that are not inherently discrete.

Figure 7.7 illustrates a fuzzy set classification scheme for wetlands,

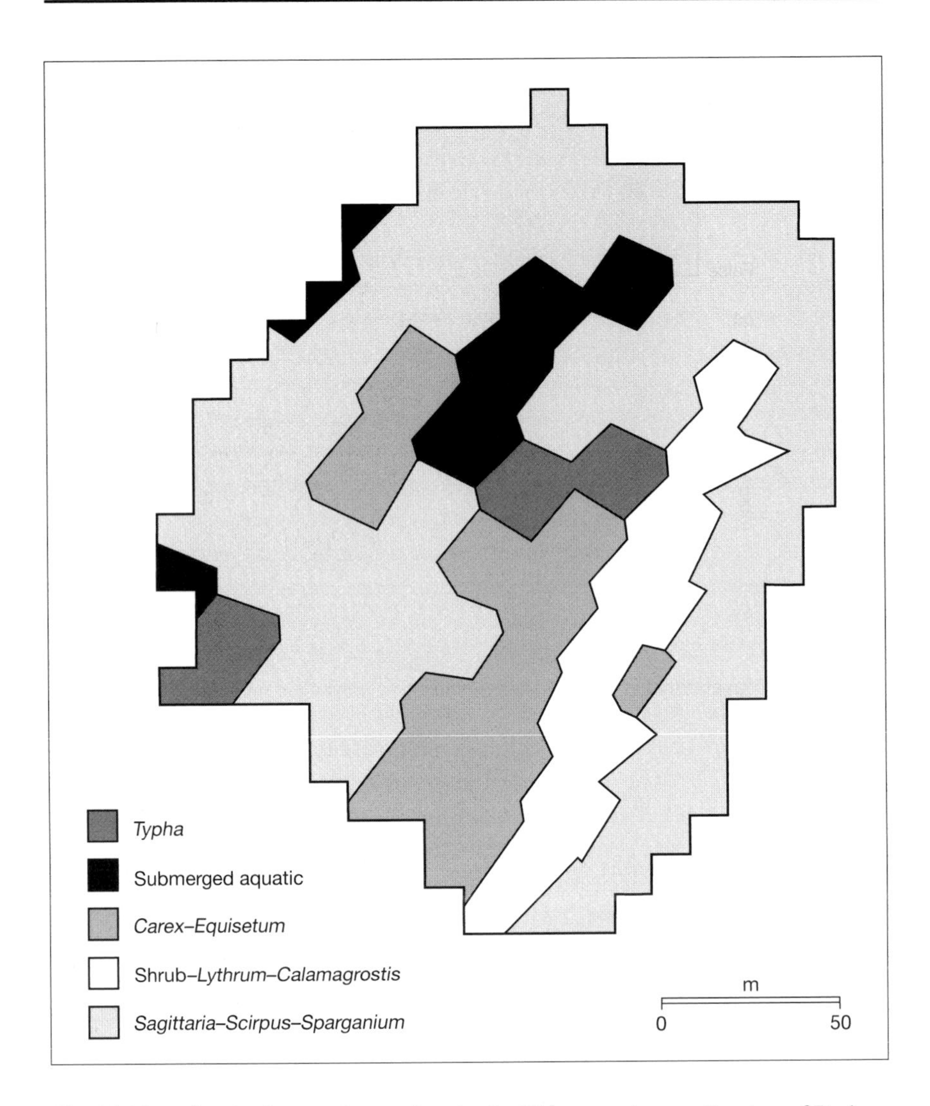

Fig. 7.6 Map of wetland vegetation produced using Thiessen polygons. Courtesy of Paul Meysembourg, Natural Resources Research Institute, University of Minnesota, Duluth.

using water depth criteria from several international definitions. Under classical set theory, any area with a water table within 30 cm below to 2 m above the surface constitutes a wetland, with areas outside those thresholds being upland or deep water, respectively. Under fuzzy set theory, the threshold boundary between upland and wetland at 30 cm is replaced by a graded boundary, such that areas with water table depth between 46 and 15 cm of the surface have a 0–100% probability, membership grade between 0 and 1, of being a wetland. Similarly, areas with

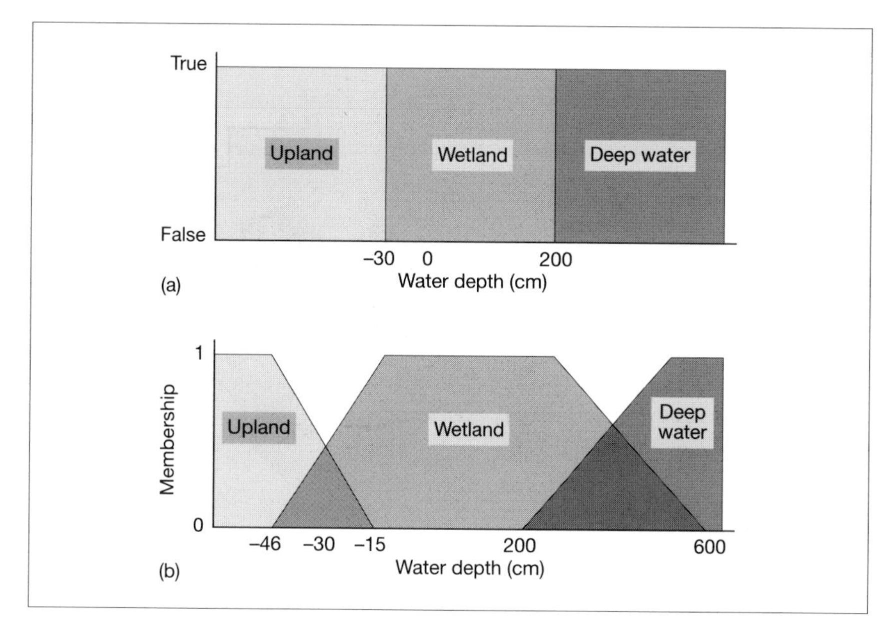

Fig. 7.7 (a) Conventional versus (b) fuzzy classification of wetland boundaries.

2–6 m of water overlying the surface also have a membership grade between 0 and 1.

Fuzzy set theory has been applied in remote sensing image analysis (Wang 1990), and was used with airborne Thematic Mapper imagery to classify heath vegetation in the U.K. (Foody 1992). Fuzzy representation has potential for GIS use with attribute data (Burrough 1989; Sui 1990; Kollias & Voliotis 1991; Leung *et al.* 1992) as well as positional data (Brimicombe 1993). Boolean algebra operations such as intersection, union, and negation can be extended easily to fuzzy set operations (Kandel 1985).

7.2 Interpolation of continuous data

In contrast to interpolation by boundary determination, the following methods interpolate data across a mathematically definable surface. Trend surface analysis and Fourier series are whole area interpolation methods, in which interpolation is based on all points in the study area. Moving averages, splines, and kriging are local interpolation methods, in which values are estimated from neighboring points only (Burrough 1986). Local interpolators allow modification of a portion of a data surface without recomputing the entire equation, which is not possible with whole-area interpolators.

7.2.1 *Whole-area interpolation*

Trend surface analysis uses least-squares regression to develop an equation that relates point locations with their attributes (Burrough 1986), assuming that each point is a sample of the same type of distribution, but that the average of the distribution can change from place to place (Clark 1980). As an example, let us assume that a Global Positioning System (GPS – see Chapter 8) has been used to measure elevation and UTM location at various points along a north-trending road on a hillslope. A scatter plot of elevation versus northing coordinate indicates that a linear equation can be fit to the data collected, of the form:

$$Z = b_0 + b_1 X \tag{7.2}$$

where b_1 is the slope of the hill and b_0 is the road's hypothetical elevation at the equator. Proceeding further, we discover that the road crests the hill and descends into and out of a swale. This new configuration is no longer suitable for linear regression, but requires a quadratic equation or higher order polynomial.

In two-dimensional trend surface analysis, an equation must be developed that relates attribute data to change in location in both the x and y direction. For a non-horizontal planar surface, the equation takes the form:

$$Z = b_0 + b_1 X + b_2 Y \tag{7.3}$$

More complex equations are used to describe curved surfaces. Trend surface analysis is used primarily for interpolating data in which there is a regular change in the attribute over two-dimensional space (e.g. groundwater elevation), or for removing broad features of a data surface (e.g. strike and dip of bedrock planes) prior to using a local interpolator to analyze variation from the broad trend.

A Fourier series describes a data surface as a linear combination of sine and cosine waves, and is useful for describing features with periodic undulations such as ripples and sand dunes (Burrough 1986). In general, however, this technique has limited utility for spatially distributed ecological data.

7.2.2 *Local interpolation*

7.2.2.1 Splines

Splines are piecewise functions fitted to a small number of data points, such that the joins between different parts of the curve are continuous (Burrough 1986). They are analogous to the flexible rulers used in drafting to create smoothed lines from a series of points. Splines are used to generate smoothed lines or polygons from data points representing categorical entities (e.g. streams, patch boundaries), and to fit isopleths (contours) to continuous data. Splines for three-dimensional surfaces are termed bicubic splines (Burrough 1986).

Splines are computed as polynomials connected at break points, which may be chosen to coincide with or fall between actual data points. A fitting tolerance can be invoked to specify how close splines must be to the data points used to compute them (Dubrule 1983). Smoothing is increased by increasing the spline "tension." Excessive smoothing can alter area and perimeter measurements when splines are used as polygon boundaries.

When using splines to construct isopleths, interval values must be chosen first. The known data values must then be interpolated to estimate the location of points along each isopleth, to which splines are fit. Different interpolation results may be produced from the same set of data points (Fig. 7.8). Some contouring packages allow break lines, such as stream traces and watershed divides, to be used to guide contouring where the source data are ambiguous.

Advantages of splines for spatial interpolation are that they retain localized features and can be computed quickly. Their primary disadvantage is that there are no direct estimates of the errors associated with spline interpolation (Burrough 1986).

7.2.2.2 Moving averages

Moving averages are used commonly to estimate data values between sampled locations. Values are computed for a local neighborhood using a two- or three-dimensional window, and assigned to a point or raster in the window center. For regularly spaced data along a transect, the moving average Z is computed for point x in the center of the symmetrical window using the Gaussian maximum likelihood estimator:

$$\hat{Z}(x) = \frac{1}{n} \sum_{i=1}^{n} Z(x_i) \tag{7.4}$$

In two dimensions, x_i is replaced by the coordinate vector X_i (Burrough 1986). Various window shapes and dimensions may be used (see Chapter 3). A larger window will tend to smooth the data more than a smaller window would, and an elongated shape will emphasize variation parallel to its major axis. Moving averages may be weighted as a function of the distance to the sample points within the window. Moving averages are susceptible to clustering of data points (Ripley 1981), a problem which is overcome by optimal interpolation methods using spatial covariance (geostatistics).

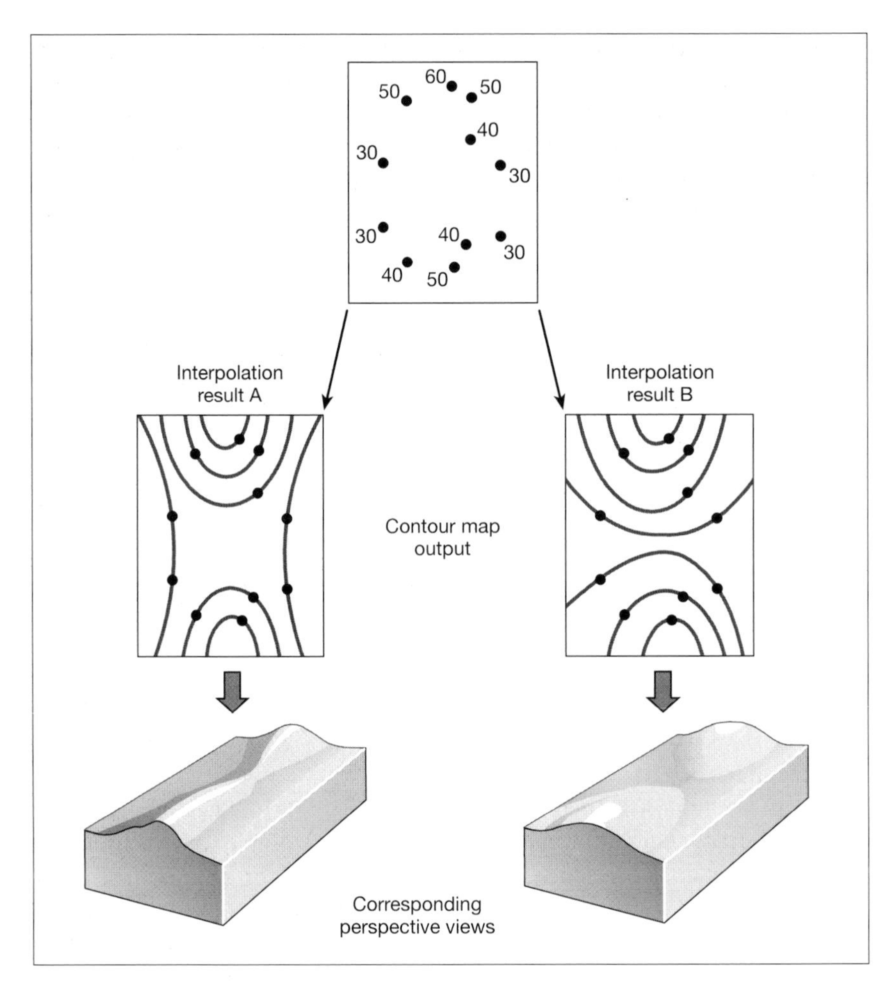

Fig. 7.8 Different interpolation results produced from the same set of data points. Interpolation result A defines a topographic saddle, whereas result B defines a valley between two hills. Adapted from *Geographic Information Systems: A Management Perspective* by Aronoff (1989) WDL Publications, Ottawa, Canada.

7.2.2.3 Spatial autocorrelation

Rather than assuming a response to be distributed uniformly in space, spatial autocorrelation tests whether the observed value of a variable at one locality is dependent significantly on values of the variable at other localities (Sokal & Oden 1978). Positions of strong similarity (or dissimilarity) are found by comparing each point with every other point (Cliff *et al.* 1975). This approach differs from traditional statistical techniques in that the location as well as the attributes of data points are taken into consideration.

Autocorrelation impairs the ability to perform standard statistical tests of hypotheses. For example, when two variables are both positively autocorrelated, the confidence interval estimated by the classical procedure around a Pearson correlation coefficient is narrower than it is when calculated correctly (i.e. taking autocorrelation into consideration), so that one declares too often that the coefficient is significantly different from zero (Legendre 1993).

The two most commonly used measures for spatial autocorrelation are Moran's I statistic and Geary's c statistic, both of which indicate the degree of spatial autocorrelation summarized for the entire data set (Cliff & Ord 1973; Anselin 1993). Moran's I is based on cross-products to measure value association, and is calculated for N observations on a variable x at locations i, j as:

$$I = (N/S_0)\Sigma_i\Sigma_j w_{ij}(x_i - \mu)(x_j - \mu)/\Sigma_i(x_i - \mu)^2 \tag{7.5}$$

where μ is the mean of the x variable, w_{ij} are the elements of the spatial weights matrix, and S_0 is the sum of the elements of the weights matrix: $S_0 = \Sigma_{ij}w_{ij}$. Geary's c statistic is expressed in the same notation as:

$$c = (N - 1)/2S_0[\Sigma_i\Sigma_j w_{ij}(x_i - x_j)^2/\Sigma_i(x_i - \mu)^2] \tag{7.6}$$

Positive spatial autocorrelation is indicated by a value of Moran's I that is larger than its theoretical mean of $-1/(N - 1)$, or a value of Geary's c smaller than its mean of 1.

Tests for spatial autocorrelation can be used to identify the extent of areas of homogeneity, but are not in themselves suitable for extrapolating over long distances, because they fail to account for regionalized variables that are too irregular to be modeled by a smooth mathematical function. This capability was provided by the work of D.G. Krige (1966) on two-dimensional weighted moving average trend surfaces and Georges Matheron (1971) on the theory of regionalized variables, which formed the basis for the contemporary field of geostatistics.

7.2.2.4 Geostatistics

Geostatistics provides a means of quantifying the effect of location on sample variability. Originally developed for ore body estimation (Krige 1966; Journel & Huijbregts 1978), geostatistics can be applied to any spatially distributed data set. The method recognizes that the spatial variation of a regionalized variable (e.g. soil, hydrology, biota) is too irregular to be modeled by smooth mathematical function, but can be described better by a stochastic surface (Burrough 1986). Interpolation weights are chosen to provide a Best Linear Unbiased Estimate of the values at unknown points, based upon the values of surrounding known points.

Geostatistics employs a regression method (**semivariogram modeling**) that assumes that the spatial variation of any variable can be expressed as the sum of three major components: (i) a structural component, associated with a constant mean value or trend; (ii) a random, spatially correlated component; and (iii) a random noise or residual error term (Burrough 1986). Geostatistics also employs an interpolation method (**kriging**) that uses the semivariogram model to generate a trend surface map that is based on sampled points and incorporates the spatial variability of the phenomenon of interest.

Geostatistics is based on the premise that points close together normally will be more related than points farther apart. It assumes that the distribution of the differences between pairs of sample points is the same over the entire area, depending only on the distance between and the orientation of the points. In other words, differences must be consistent, **not** constant, across space (Clark 1980). Given this assumption, known as the intrinsic hypothesis, then the variance between data points depends only on the distance and direction that separates them, and not on their absolute location. The variance, 2γ, at points separated by a distance vector $\boldsymbol{h}$ is a measure of the influence of the samples over neighboring areas within the sampled domain, and is estimated from the values of variable Z at locations x_i as:

$$2\hat{\gamma}(\boldsymbol{h}) = \frac{1}{\boldsymbol{h}} \sum_{i=1}^{n} [Z(x_i) - Z(x_i + \boldsymbol{h})]^2 \tag{7.7}$$

where n is the number of pairs of sample points. The **semivariance**, $\gamma(\boldsymbol{h})$, is equal to one-half of the variance.

SEMIVARIOGRAM MODELING. A semivariogram is a plot of semivariance as a function of distance vector $\boldsymbol{h}$ between the samples. All possible sample pairs are grouped into classes (**lags**) of approximately equal distance. Choice of the lag interval is a trial and error process, with the objective of obtaining the maximum detail at small distances (i.e. small lags) without being misled by structural artifacts or "noise" due to the particular class interval used (Englund & Sparks 1988). There should be an adequate number of pairs to estimate the semivariogram value for each lag, usually 50 or more. The highest lag distance typically used for model development is 50–80% of the maximum pair distance. Figure 7.9 illustrates the influence of different lag spacings on data representing soil organic matter.

Once the semivariogram has been calculated, a mathematical model is fitted to the data points either visually or by parametric methods (Cressie 1991). The two major groups of semivariogram models are those that reach a plateau (called a **sill**) and those that do not. Of the models

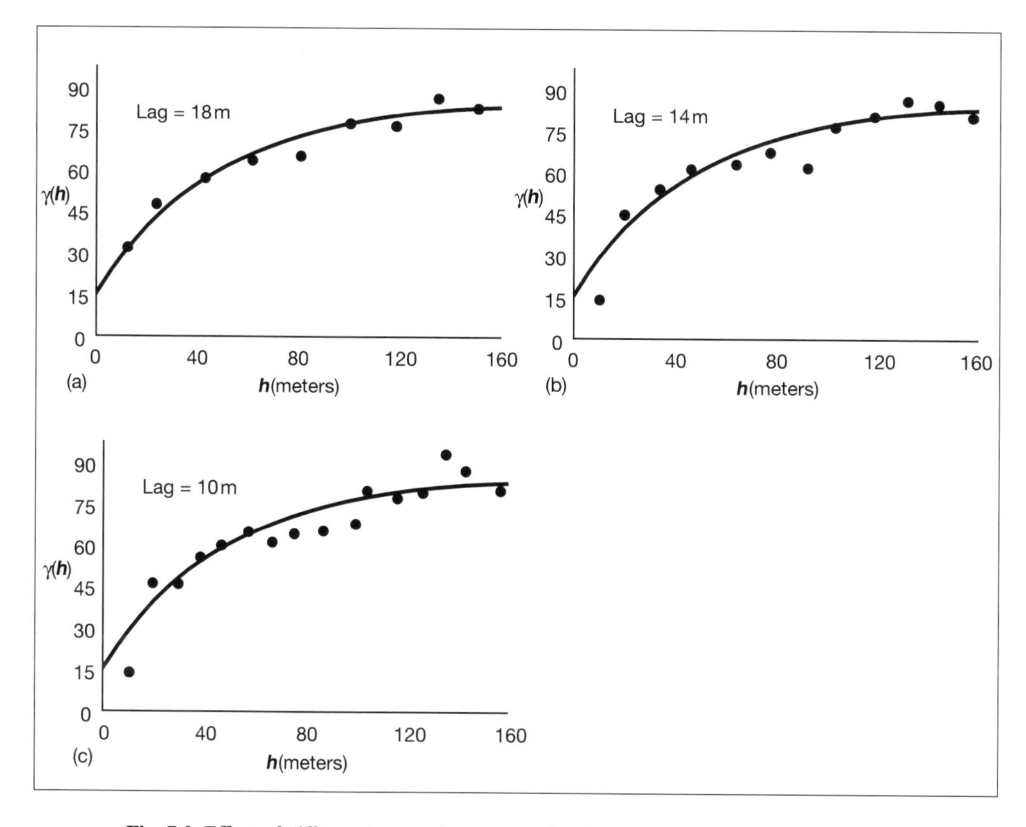

Fig. 7.9 Effect of different lag spacings on semivariogram values. $\gamma(\boldsymbol{h})$ = semivariance, $\boldsymbol{h}$ = distance.

without a sill, the most frequently used is the **linear** model, a straight line with a positive slope. Several models with sills are illustrated in Fig. 7.10. A **spherical** model rises rapidly, then curves to the sill. An **exponential** model rises more gradually, and is asymptotic to the sill. A **Gaussian** model is parabolic near the origin, then rises steeply toward the sill

Although all of the hypothetical semivariogram curves shown in Fig. 7.10 intersect the *y*-axis at the origin, it is quite common for a model fitted to real data to intersect the *y*-axis at a positive value of $\gamma(\boldsymbol{h})$ rather than the origin (e.g. Fig. 7.9). This value is known as the **nugget**. The nugget combines the residual variations of measurement errors together with spatial variations at distances shorter than the sample spacing (Burrough 1986). In some cases, the nugget is so large that the experimental semivariances show no tendency to increase with the lag, and the value of the sill is equal to the value of the nugget. A **nugget effect** model thus represents purely random (i.e. not spatially autocorrelated) behavior.

A useful characteristic of models with sills is that the constant semi-

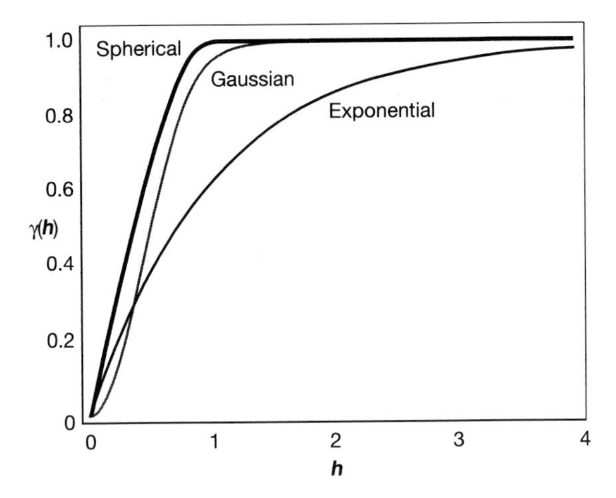

Fig. 7.10 Semivariogram models with sills. $\gamma(\boldsymbol{h})$ = semivariance, $\boldsymbol{h}$ = distance.

variance attained at the sill is equal to the ordinary variance of the sampled entity. The distance at which the model reaches the sill is called the **range**, and is the distance at which data values are no longer spatially autocorrelated. The nugget semivariance, the sill semivariance, and the range distance are used to fit sill models (Fig. 7.11).

Although the previous examples focus on the construction of semivariograms for samples with spatial autocorrelation, geostatistics can also be used to model **temporal autocorrelation**. Robertson (1987) provides an example of semivariogram construction for *Rhodomonas* sp. density in the epilimnion of a southwest Michigan lake from August 1982 to August 1983, in which the lag intervals are in numbers of days rather than

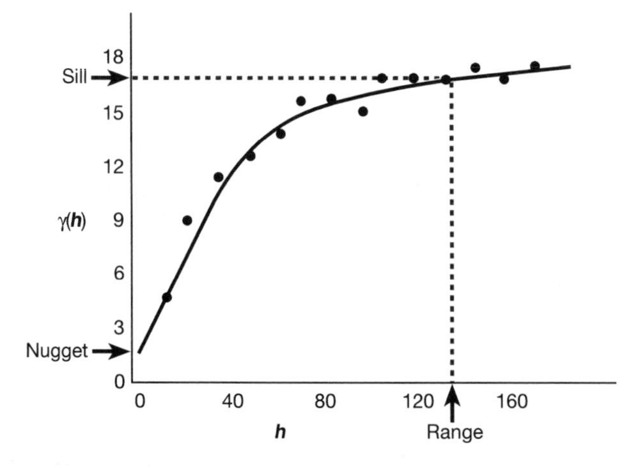

Fig. 7.11 Components of a semivariogram model with a sill. $\gamma(\boldsymbol{h})$ = semivariance, $\boldsymbol{h}$ = distance.

distance. These variance estimates can be useful for judging whether temporal patterns of epilimnetic *Rhodomonas* densities differ significantly from patterns for other species or from temporal patterns of environmental variates (Robertson 1987).

Any linear combination of valid semivariogram models is also a valid semivariogram, so that complex models can be developed using **nested** structures. If, for example, the semivariogram is parabolic near the origin but fails to reach a sill then a combination of Gaussian and linear models might be appropriate. Nested structures are first determined by generating an overall model for the semivariogram, then adding a second structure such that the sum of the sills for each nested structure equals the sill of the overall model.

ANISOTROPIC MODELING. The semivariogram analysis discussed so far has been **omnidirectional**, modeling spatial variation equally in all directions. However, it is quite common for natural features to be elongated, so that the variation in one direction is greater than that in another. For example, points that are downslope or upslope from one another may be differently autocorrelated than points located along the same contour interval (Robertson 1987). Where such **anisotropy** exists, a directional component must be included in the semivariogram model. Directional semivariances are computed for points that fall within a tolerance angle (generally 10–25%) centered around a given compass direction (Fig. 7.12).

Analysis of anisotropic semivariograms is illustrated by Fig. 7.13 for

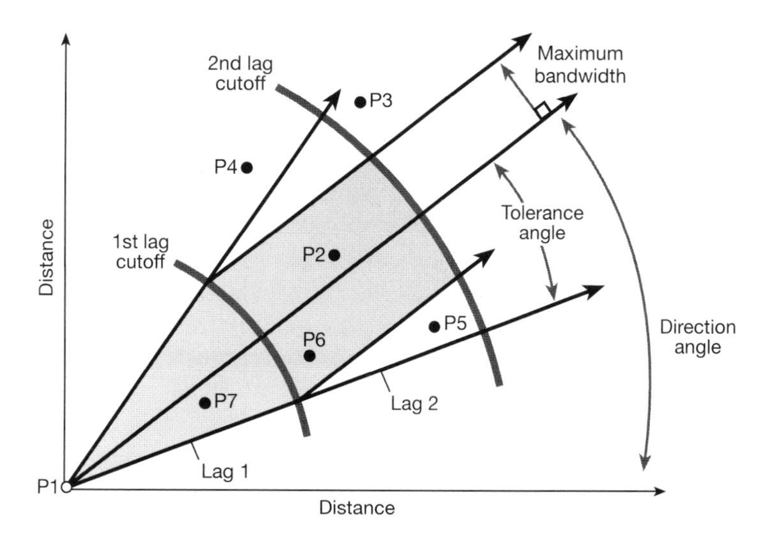

Fig. 7.12 Scatter plot of sample points illustrating selection of points for computation of directional semivariance in relation to point P1. From Englund and Sparks 1988.

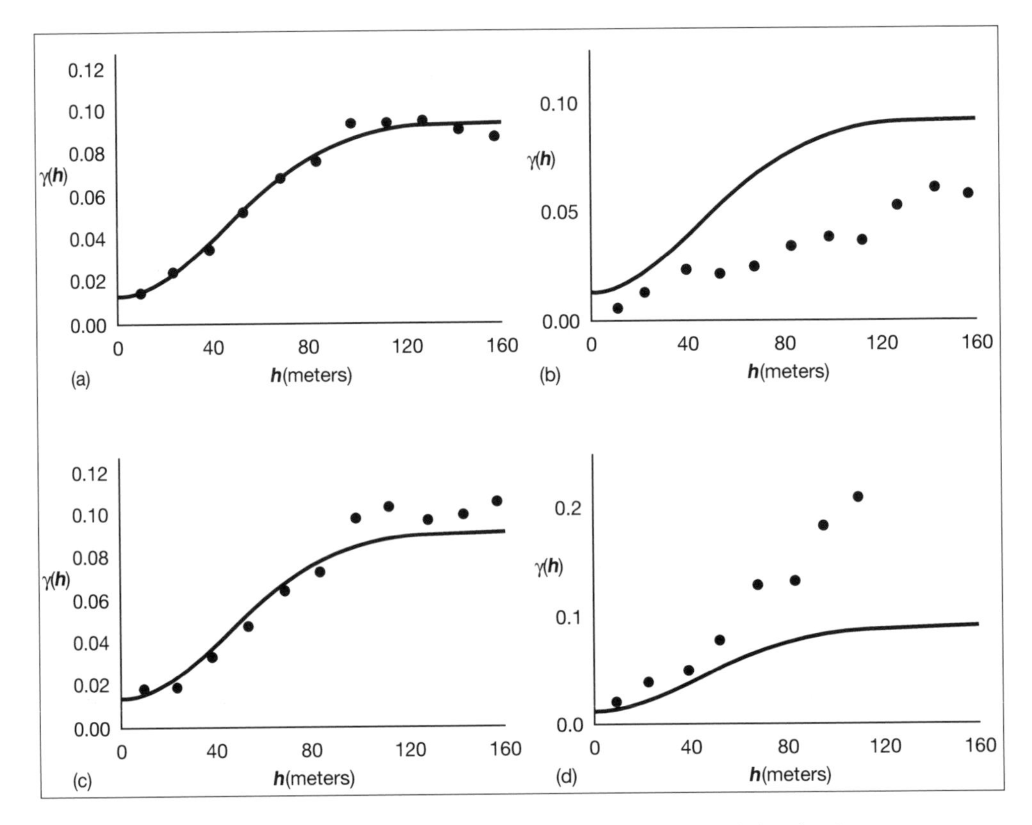

Fig. 7.13 Semivariograms of soil bulk density in a riverine wetland: (a) omnidirectional, (b) 0° direction, (c) 45° direction, (d) 90° direction.

soil bulk density (Mg m^{-3}) in a riverine wetland, where the 0° direction parallels the river. The omnidirectional model for this data set, plotted on all four semivariograms, has a Gaussian shape, a nugget of 0.013, a sill minus nugget value of 0.08, and a range of 110 m (Fig. 7.13a). Values of $\gamma(\boldsymbol{h})$ for the 0° directional semivariogram fall below the omnidirectional model, indicating that the range should be much longer than 110 m (i.e. the distance at which the data values are no longer spatially autocorrelated is much farther away). Conversely, semivariogram values in the 90° direction, perpendicular to the flow of the river, fall above the omnidirectional model, indicating that the range should be shorter than 110 m (i.e. the distance at which the data values are no longer spatially autocorrelated is much closer). The 45° directional semivariogram fits the omnidirectional model fairly well, because semivariance values are intermediate between those for the 0° and 90° directions. Based on this analysis, a directional semivariogram was developed, with the major and minor ranges in the 0° and 90° directions, respectively.

KRIGING. Once a semivariogram model has been developed, kriging can be used for the estimation of values at unsampled locations. Kriging is defined as "a weighted moving average interpolation method where the set of weights assigned to samples minimizes the estimation variance, which is computed as a function of the variogram model and locations of the samples relative to each other" (Englund & Sparks 1988). The kriging predictor weights are chosen in an optimal way, optimal in the sense that they provide a Best Linear Unbiased Estimate of the value of the variable at a given point. Englund and Sparks (1988) describe several advantages of kriging over other interpolation methods:

- **Smoothing** — kriging smoothes estimates based on the proportion of total sample variance accounted for by random "noise." The noisier the data, the less individual samples represent their immediate vicinity, and the more they are smoothed.
- **Declustering** — the kriging weight assigned to a sample is lowered to the degree that its information is duplicated by nearby, highly correlated samples. This helps mitigate the impact of oversampling "hot spots."
- **Anisotropy** — when samples are more highly correlated in a particular direction, kriging weights will be greater for samples in that direction.
- **Precision** — given a variogram representative of the area to be estimated, kriging will compute the most precise estimates possible from the available data.

In addition, kriging can be used to design an optimal strategy for sampling an area, knowing only something about the degree of sample interdependence (the shape of the semivariogram) for samples within the interpolation domain (Burgess *et al.* 1981; McBratney & Webster 1983; Russo 1984; Webster & Burgess 1984; Atkinson & Harrison 1993). Where samples are strongly autocorrelated over small sampling intervals, pre-sample kriging can show where to add sample points to bring estimation precision to a desirable level (Robertson 1987).

Punctual kriging is an exact interpolator, so that estimated values are identical to measured values where interpolated points coincide with sample locations (Delhomme 1978; Burgess & Webster 1980a). **Block kriging** is more appropriate when the sampled data are meant to represent a larger area, or when local discontinuities might obscure longer range trends (Burgess & Webster 1980b). Kriging, whether punctual or block kriging, generates a regularly spaced data grid. This grid can either be brought directly into a raster GIS or can be used to generate isopleths for a vector GIS. Figure 7.14 is an isopleth map of soil bulk density, kriged from the anisotropic semivariogram model developed above.

Geostatistical applications in ecology are still uncommon (Robertson 1987; Michener *et al.* 1992; Rossi *et al.* 1992; Johnston 1993; Legendre 1993), but geostatistics has been used widely to quantify the spatial

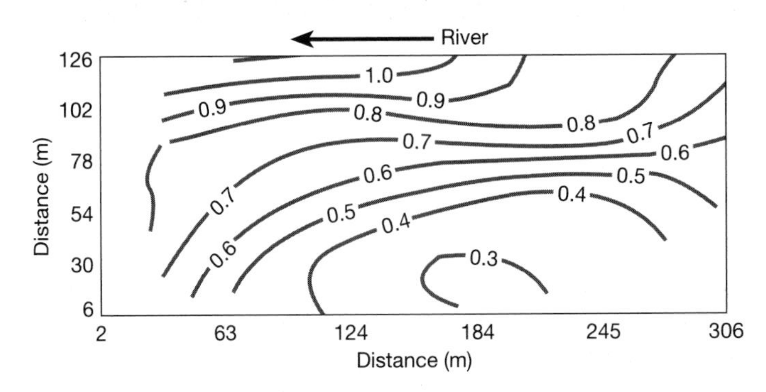

Fig. 7.14 Contour plot of soil bulk density (Mg m^{-3}) in a riverine wetland developed by kriging of anisotropic model developed from semivariograms shown in Fig. 7.13.

variability of soil properties: soil moisture (Bell *et al.* 1980; Yates & Warrick 1987), hydraulic properties (Russo & Bresler 1981; Sisson & Wierenga 1981; Vieira *et al.* 1981), temperature (Hatfield *et al.* 1982; Vauclin *et al.* 1982; Yates *et al.* 1988), and other characteristics (Gajem *et al.* 1983; Trangmar *et al.* 1987). Geostatistical techniques have also been used to analyze the spatial distribution of nitrogen transformation processes, such as denitrification, nitrification, and nitrogen mineralization (Folorunso & Rolston 1985; Robertson *et al.* 1988).

Some GIS are beginning to include geostatistical algorithms for interpolating point data. However, there are several excellent inexpensive geostatistical packages currently available, including GEO–EAS (Englund & Sparks 1988), GEOPACK (Yates & Yates 1990), and GSLIB (Deutsch & Journel 1992). A number of spatial analysis programs which provide capabilities not available in general-purpose statistical packages are described by Legendre (1993).

CHAPTER 8

Global Positioning Systems and GIS

8.1 Introduction

Knowing one's location on Earth has been a problem that has plagued humans since they first started exploring their habitat. The challenge of determining location has expanded the frontiers of science with innovative solutions such as the compass and celestial navigation. Field ecologists have been hampered by the lack of a simple, inexpensive means of determining their location relative to the rest of the Earth, a barrier that has discouraged location-specific research.

At spatial scales of centimeters to tens of meters, sample locations can be plotted using compass and measuring tapes. At longer distances, surveying with rod and transit or laser devices can be used for stationary points, and radiotelemetry can be used for mobile objects, such as tracking wildlife. These methods are all time-consuming, and the latter requires expensive equipment and expertise. Furthermore, unless measured sample points have been referenced to a known geographic location, such as a U.S. Geological Survey (USGS) benchmark, they are not **georeferenced**, or related to Earth coordinates. Lack of georeferencing prevents field-sampled data from being used with independent sources of environmental data, such as soil or topographic maps, in a GIS.

The Navstar Global Positioning System or **GPS** is an innovative solution to this age-old problem. Navstar consists of a constellation of satellites orbiting the Earth that continuously broadcast location information (Fig. 8.1). This information can be intercepted by GPS receivers to identify location in Earth coordinates within meters to millimeters. The system was developed at a cost of over $10 billion by the U.S. Department of Defense (DoD) for military applications, but is also available for civilian use.

This chapter discusses the basics of GPS as they pertain to ecological applications. A more comprehensive, yet easily understood, discussion of GPS theory and operation is contained in *GPS, A Guide to the Next Utility* (Hurn 1989). Examples of GPS applications in ecology and other fields are published in the trade journal *GPS World.**

* For subscription information, write to *GPS World*, P.O. Box 6148, Duluth, MN 55806-6148, USA.

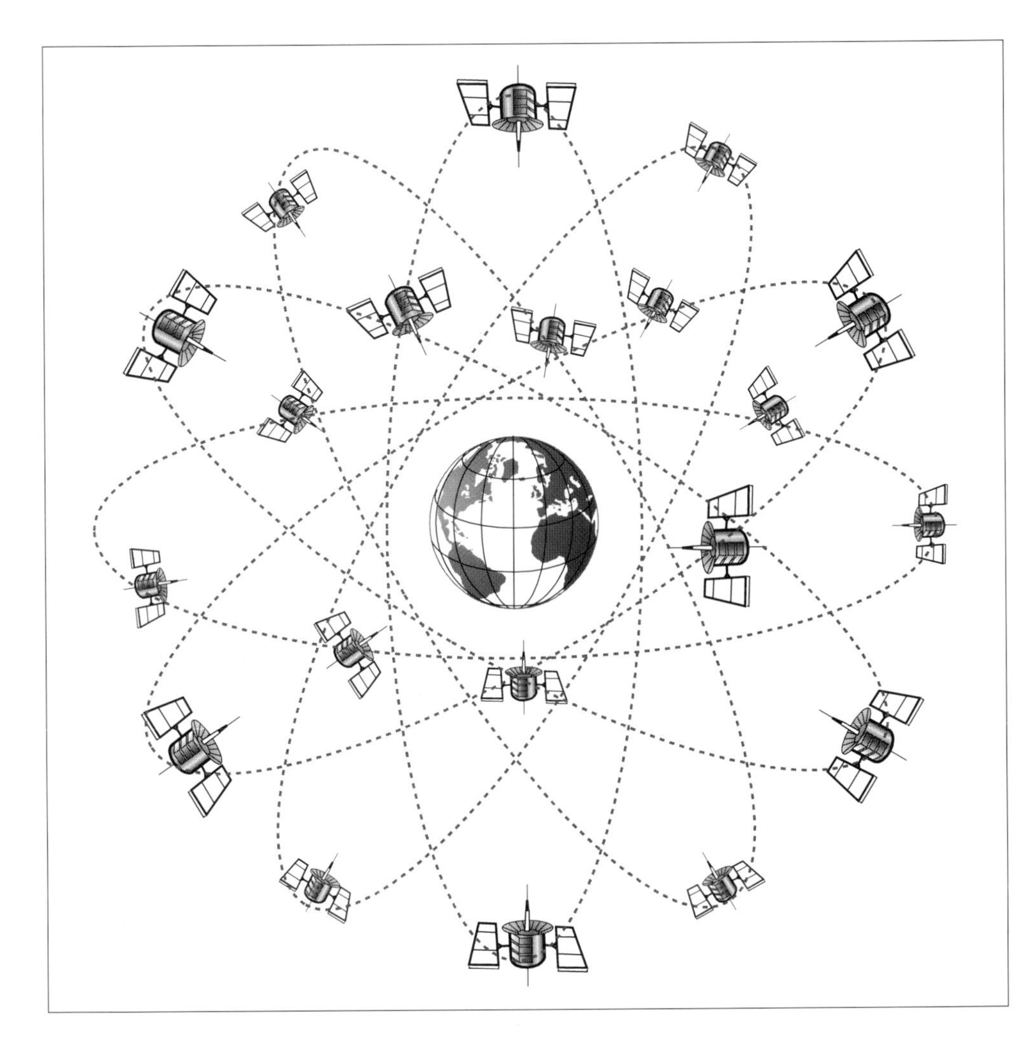

Fig. 8.1 GPS satellites orbiting the Earth.

8.1.1 How does it work?

GPS uses the geometric principle of **triangulation** to determine location. In traditional surveying, triangulation is done by measuring distances on the ground, but in GPS surveying, distances are measured from the satellites by means of radio signals. One measurement narrows location in the universe down to a sphere with radius equal to the distance from the satellite (Fig. 8.2a). Measuring the distance to two satellites narrows location even further to the circle that falls at the intersection of the two spheres defined by those distance radii (Fig. 8.2b). Measuring the distance to a third satellite limits the location to one of two points in space where the third sphere intersects the circle. The trigonometric solution

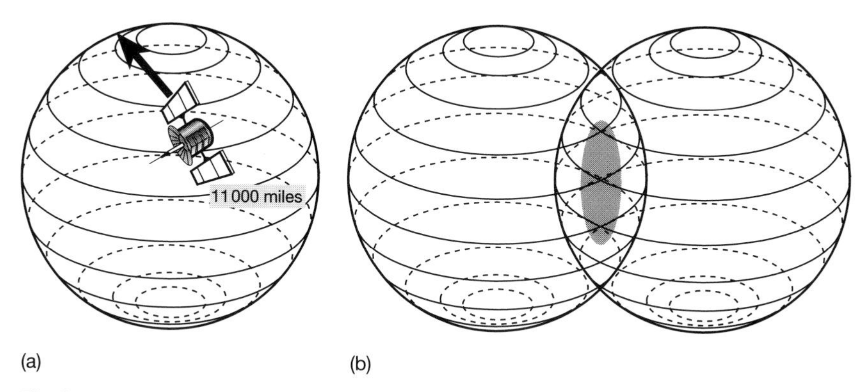

Fig. 8.2 Determining location in space based on two GPS readings: (a) a single distance defines a sphere around a satellite, (b) the intersection of two spheres defines a circle on which the location may occur.

would require a fourth measurement to determine which of these two points is correct, but one of the two points usually can be rejected because it does not lie near the surface of the Earth, occurring deep within the Earth or out in space. Therefore, a minimum of three perfect distance measurements is theoretically required to determine an Earth location in three dimensions.

When the elevation of a location is known, such as sea level, the trigonometric solution is even simpler, because the "sphere" of the Earth can be substituted for one of the measurements. Operation of a GPS at a known elevation is termed **two-dimensional operation**, and requires only two perfect distance measurements.

A GPS measures distance through the use of radio signals and very precise clocks. Both the GPS satellite and receiver generate a complicated set of pseudorandom codes at exactly the same time (Fig. 8.3). The satellite sends this signal to the receiver, which compares the received

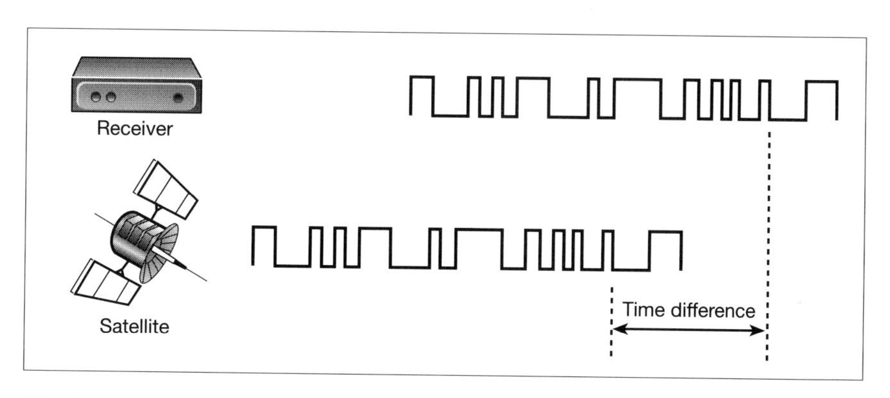

Fig. 8.3 The time lag between pseudorandom codes generated by the satellite and the receiver is used to measure distance.

signal to the one it just generated, determining the time lag between them. This time lag is multiplied by the speed of the radio signal (299 460 km/s, the speed of light) to determine the distance between the satellite and the receiver.

Accurate measurement of the time lag between pseudorandom codes is essential to GPS functioning. Therefore, the satellites use very precise atomic clocks to determine the time at which the signal was sent, and most receivers use clocks with nanosecond (10^{-9}) accuracy. Even so, small errors in timing measurement can result in large errors in distance calculated because radio signals travel so fast. The solution to this problem lies in taking an additional satellite distance measurement, illustrated by Fig. 8.4. When the clock in the GPS receiver is slow, then the distance measured will be too long, and the locations determined from three different satellites will not intersect at the same point. To correct for clock error, a GPS receiver automatically checks for consistency of locations determined by different satellites, and computes the offset by algebra-

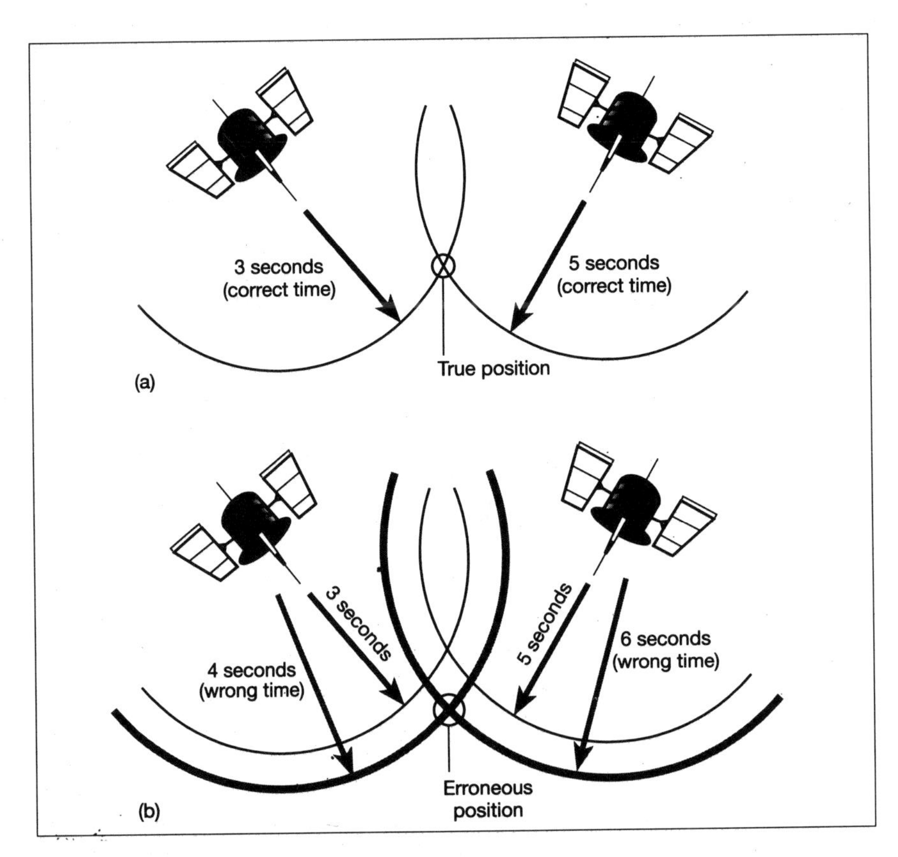

Fig. 8.4 Location error induced by time measurement imperfections: (a) correct distance measurements define true position, (b) incorrect measurements yield erroneous position.

ically solving for the location. This same principle is used in three dimensions (e.g. when elevation is not known), but requires readings from four satellites (Fig. 8.5).

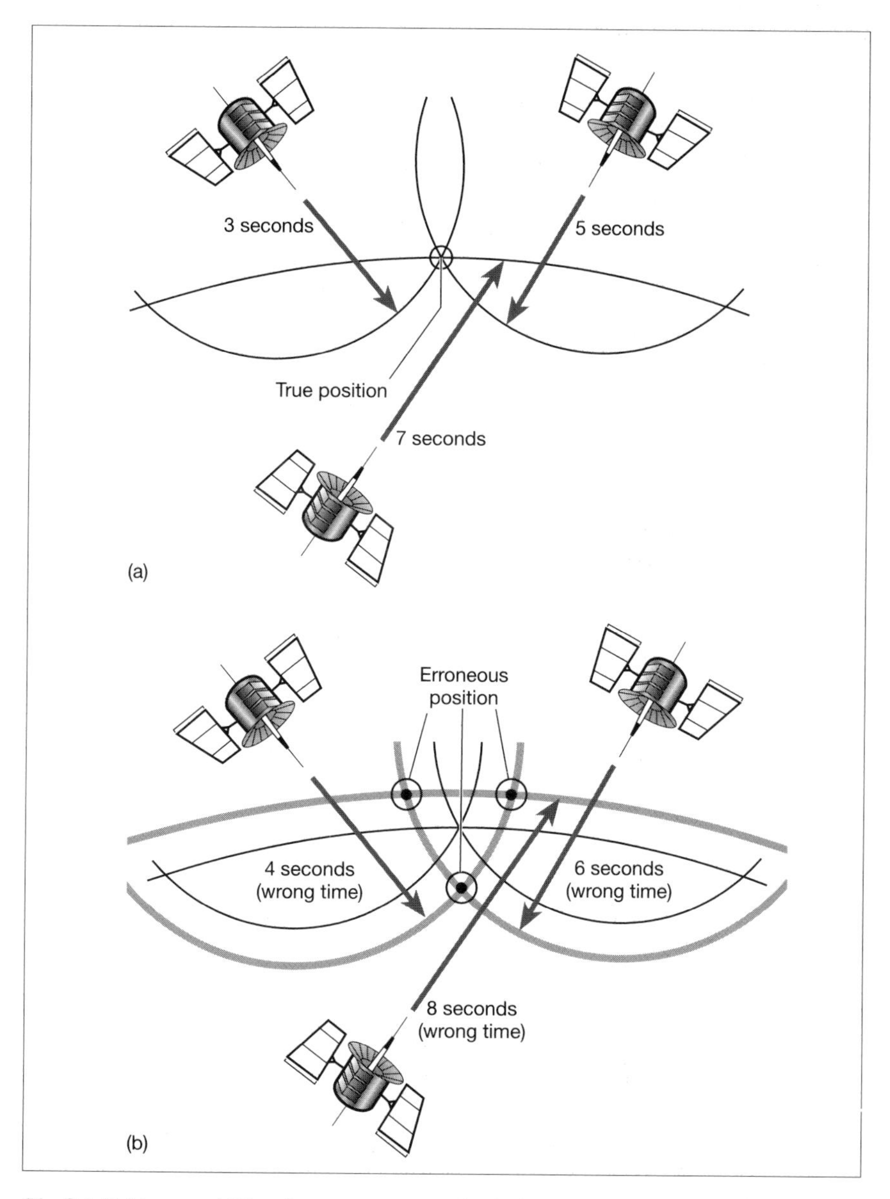

Fig. 8.5 Taking an additional measurement to check clock accuracy: (a) only one point lies at the intersection of three true distance measurements, (b) a slow internal clock yields three different positions for the same point, thereby alerting the GPS that its internal clock is off.

8.1.2 Differential correction

Out of concern for national security, the U.S. DoD frequently degrades the pseudorandom code using **selective availability** (S/A). When S/A is activated, it is the largest component of GPS error, greatly decreasing locational accuracy. Unlike timing errors, which can be corrected simply by taking an additional satellite measurement, correction of S/A error requires the use of two GPS receivers operating simultaneously through a process called **differential correction**. Under differential correction, a reference receiver is positioned at a known location. The reference receiver can be a community base station dedicated to continuously recording GPS data, or it can be another GPS positioned over a USGS benchmark or other surveyed site. Because the reference receiver's location is known, its readings can be used to compute the error induced by S/A and other sources (e.g. receiver clocks, satellite position, atmospheric delays). The error occurring at the reference GPS will be virtually identical to the error affecting GPS readings at the other receiver.

Correction of the readings taken by the roving GPS can be either instantaneous (real-time differential) or after the fact (post-processing differential). In **real-time differential** mode, correction factors are sent by radiotelemetry from the reference receiver to the roving receiver, so that readings can be corrected as they are taken. In **post-processing differential** mode, the readings are taken and later corrected using computer software and a data file downloaded from the base station. The base station can be located hundreds of kilometers from the roving GPS unit, but must have received simultaneous data from at least four of the same satellites as the roving unit. An advantage of real-time differential processing is that problems with the reference receiver are immediately known, whereas with post-processing differential problems may not become evident until after the field data have been collected, often at great time and expense.

Due to the increasing use of GPS to locate delivery trucks and other moving vehicles, a differential correction signal using a standard format promulgated by the Radio Technical Commission for Maritime Services **(RTCM)** is broadcast over FM radio frequencies in many metropolitan areas. Users subscribe to this service for a fee, and use an interface to automatically correct for S/A in real time. Although the Radio Technical Commission for Maritime Services established the format for RTCM, its use is not restricted to maritime environments.

8.1.3 Satellite geometry

In order to compute locations in three dimensions there must be four satellites visible to a GPS receiver. Fortuately, the full constellation of 24+ GPS satellites makes this possible anywhere on Earth where the view of the sky is not obstructed. Each satellite orbits the Earth every 12 hours, so their configuration changes throughout the day. Although the satellites are in very precise orbits, minor variations are caused by gravitational pulls from the moon and sun and the pressure of solar radiation on the satellite, known as **ephemeris errors**. The U.S. DoD monitors each satellite's altitude, position, and speed as it passes over twice a day, and relays that information to the satellite, which in turn relays correction data to GPS receivers along with the pseudorandom code. GPS receivers can use the relayed data to correct for ephemeris error.

Certain satellite positions provide better accuracy than others. Even though the speed of light (and GPS radio signals) is constant through a vacuum, radio signals slow down slightly when passing through denser media, such as the Earth's atmosphere. Radio signals from satellites that are low on the horizon must pass through a longer distance of atmosphere, and are therefore more susceptible to error. They are also susceptible to being bounced off objects on the Earth's surface, producing the kind of **multipath error** that causes ghosts on TV images. For this reason, most GPS receivers can be programed to exclude readings from satellites with angles low to the horizon (e.g. $< 15°$).

GPS signals are blocked by obstructions such as mountains, buildings, or dense tree canopies that lie between the receiver and a satellite. Strategies for taking GPS readings in forested areas include: (i) taking readings at times when there are more satellites at high angles overhead (i.e. $> 20°$ over the horizon); (ii) waiting until deciduous leaves have fallen; and (iii) elevating the antennas above the canopy with a long pole or helium balloon tether. The latter two options may be impractical in tall, dense coniferous forests, which limits the utility of GPS for some ecological applications. In mountainous areas, software should be used to determine the times of day during which a suitable satellite configuration will be visible in the sky.

The geometry of the satellite constellation also affects accuracy. When the angle in the sky between two satellites is very small or very large, there is a larger zone of overlap between their distance measurements (Fig. 8.6a). This zone is smallest when the angles are orthogonal (Fig. 8.6b). Therefore, the satellite configuration can magnify or lessen the other uncertainties associated with GPS readings, known as Geometric Dilution of Precision (**GDOP**). Under good conditions, GDOPs range from four to six, and multiply the error from various sources (Table 8.1). Therefore, minimizing the GDOP can greatly influence the accuracy of GPS readings.

8.1.4 Satellite geometry applied to Earth geometry

Satellite orbits are expressed in Cartesian coordinates (x,y,z), but locations on the Earth's surface are traditionally expressed in curvilinear

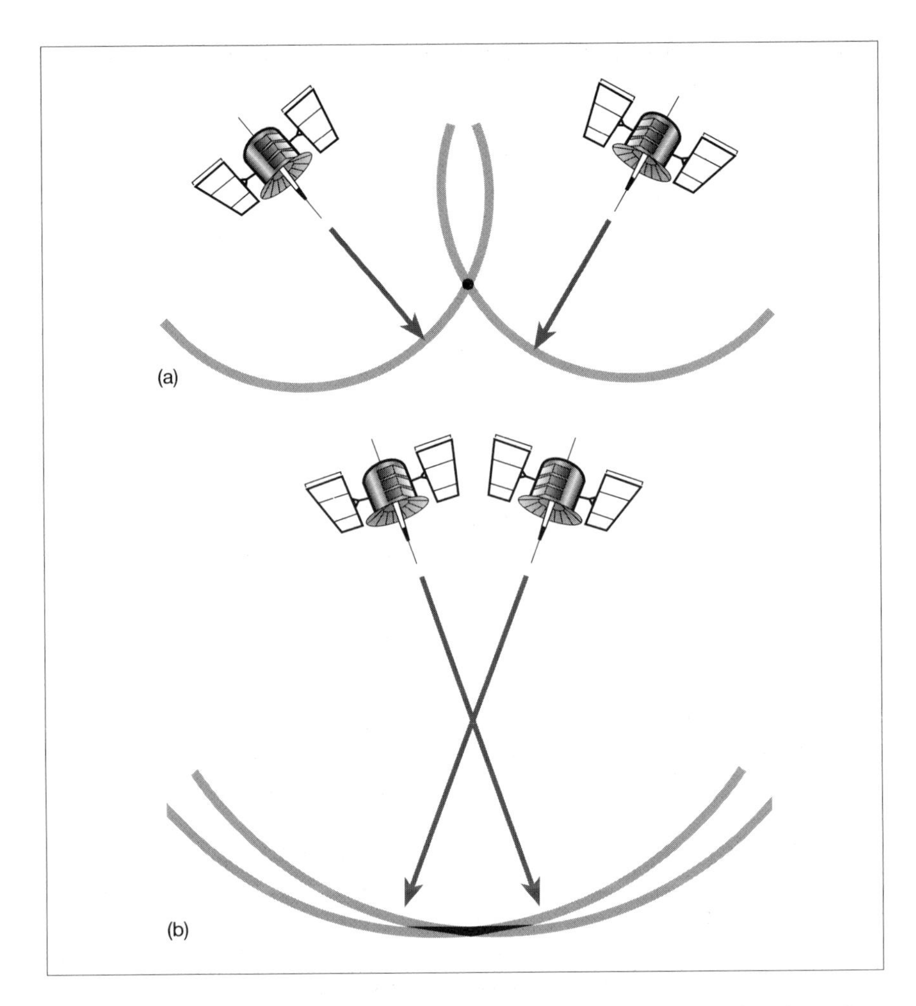

Fig. 8.6 There is a larger zone of overlap between distance measurements when the angle between two satellites is small (a) than when the angles are orthogonal (b).

geodetic coordinates: latitude, longitude, and height. A GPS measures its location in Cartesian coordinates, and must convert those coordinates to geodetic coordinates to make them meaningful for Earth-bound applications. GPS coordinates are referenced to an ellipsoid which has its origin at the center of the Earth. Unlike a geoid model of the Earth (see Section 2.4), this ellipsoid is a smooth mathematical surface and is not affected by gravity. If the height of the geoid above the ellipsoid is known, then GPS-measured ellipsoid heights can be converted to orthometric heights, elevations that use sea level (or a vertical datum that is parallel with sea level, such as NAVD-88) as their origin or zero value (Cheves 1997).With better data on gravitational anomalies and advances in GPS technology,

Table 8.1 Error budget for a typical GPS receiver using pseudorandom codes only. After Hurn 1989.

Source	Error (m)
Satellite clock error	0.6
Ephemeris error	0.6
Receiver errors	1.2
Atmosphere/ionospheric	3.7
Worst case S/A	7.6
Subtotal (root-square sum)	4–9*
Total error with GDOP of 5 and no S/A	20
Total error with GDOP of 5 and S/A implemented	45

*Depending on S/A.

height determination by GPS is expected to replace traditional leveling techniques in all but the most demanding applications (Schwartz & Sideris 1993).

8.2 Types of GPS receivers

GPS receivers have a wide range of features and costs. An industry survey conducted in 1997 by the trade magazine *GPS World* listed 394 GPS receivers from 61 manufacturers, at prices ranging from $50 to $61 500 (GPS World 1997). GPS receivers are packaged in different housings (e.g. lightweight, portable units for field use), have different antenna styles, keypads, displays, data storage and output capabilities, power supplies and power consumption rates, and may or may not come with software needed for error correction and data conversion to GIS-usable formats (Langley 1991). In addition to measuring time and location, many GPS receivers can also calculate change in location over space (i.e. direction) and over time (i.e. velocity). GPS receivers are designed for different applications: military, navigation, surveying, vehicle tracking, timing/frequency, or general position location. Some take much longer than others to get their first fix when ephemeris, satellite configuration, and initial position and time location are known. Vendor-rated positional accuracies, measured as SEP (spherical error probable – the radius of a sphere within which the reading will fall 50% of the time) are as low as a few millimeters with post-processing differential correction in the most precise units.

One difference among GPS receivers is the number of channels they use to receive satellite information. The best GPS receivers use a number of parallel channels (typically 6–12) to simultaneously and continuously take readings from the same number of satellites. This minimizes the GDOP, thereby maximizing accuracy. Sequencing receivers move one or more channels from one satellite to the next until they have gathered data from at least four (the minimum number of satellites required to calculate an accurate position in three dimensions), rather than fixing channels on a particular satellite. A sequencing receiver must therefore remain stationary for a longer time before it can accurately measure location.

Each GPS satellite transmits two carrier signals for positioning purposes: L1 at a frequency of 1575.42 MHz, and L2 at a frequency of 1227.60 MHz. Modulated into the L1 carrier are two pseudorandom codes: the C/A (Clear/Acquisition) code and the P-code, which can be encrypted for military use only. The L2 carrier is modulated by the P-code and the navigation message. Some GPS receivers can measure not only a pseudorandom code, but also its carrier phase. **Carrier phase** GPS receivers have the advantage of being able to correct for most of the common errors inherent in a GPS signal (clock error, tropospheric delay) by differencing the carrier phase measurements on L1 and/or L2, and have millimeter-level precision as compared to the meter-level precision attainable using pseudorandom codes alone (Kleusberg 1992). **Dual frequency** GPS receivers, those that can measure both the L1 and L2 carrier phase, can also remove ionospheric effects (Langley 1993).

GPS performance is improving and prices are dropping as the technology matures. Ecologists seeking information about current developments in GPS should consult *GPS World*, which publishes a comprehensive survey of receiver features and costs each January.

8.3 GPS for field studies

GPS provides ecologists with a capability that heretofore was prohibitively expensive for most research applications: knowing the georeferenced location of their field sites. This capability is particularly helpful for research in wilderness areas and open waters that lack landmarks that can be related to mapped coordinates.

The accuracy of GPS locations depends on the particular device used, a fact which ecological researchers should consider when reporting and analyzing results. Researchers in Rhode Island reported that data points collected with a Trimble Pathfinder Basic portable receiver, a three-channel model commonly used for field applications, were within 6 m of true position with differential correction, and within 73 m without (August *et al.* 1994). Initial attempts to use this same GPS model to georeference wetland sample sites (C.A. Johnston & J.P. Bonde, unpublished data) were unsuccessful because the sites were only 10–20 m apart, close to the accuracy level attainable with differential processing. Taking the average of repetitive fixes at a single location increases accuracy and precision.

GPS is advancing ecology in a wide array of applications:

Wildlife – GPS has been used to map habitats for a variety of animal species, including elephants, sea turtles, monkeys, snow leopards, and endangered birds. GPS receivers have been miniaturized for use in radio collars for tracking animal movements (Moen *et al.* 1996). GPS tracking equipment is less bulky and more accurate than that used in traditional radiotelemetry and automatically provides locations in geodetic coordinates.

Forestry – When forest landholders sell logging rights, they need to verify that the area logged does not exceed the area in the bid. The U.S. Forest Service has used GPS to determine actual boundaries of timber sale areas (Peterson 1990). This method is more accurate than conventional verification techniques, and is also more legally defensible in disputed sales.

In Finland, GPS is being used to optimize logging operations (Tolkki & Koskelo 1993). After trees are harvested, the GPS location and size of each log pile are recorded. This inventory information is entered into a GIS, which designs logging truck routes to minimize transportation costs and maximize wood quality by reducing the amount of time logs lie on the ground. Each logging truck is equipped with a mobile PC and GPS receiver, and can display location and attribute data about any log pile, as well as its own position on a digital map. The more efficient transportation has resulted in lower inventory levels and faster inventory turnaround.

Aerial reconnaissance using GPS has been used to speed forest fire evaluation and control. The entire 35-mile perimeter of the Swedlund fire in the Black Hills of South Dakota was mapped in about 55 minutes by a helicopter overflight with a GPS strapped to the boom (Drake & Luepke 1991). The information was downloaded to a GIS and used to update forest inventory maps, develop silvicultural prescriptions, and assist other fire-related management activities.

Limnology and oceanography – Until the advent of GPS technology, the Loran-C navigational system was the most common means of georeferencing sample points in coastal waters, and accurate positioning outside the range of Loran-C signals was virtually impossible. GPS has enabled researchers to document the location of sample sites in inland lakes such as Utah Lake, being surveyed for bathymetry and the location of thermal springs, and Hartwell Lake (on the South Carolina-Georgia border), being monitored for polychlorinated biphenyl (PCB) contamination (Baskin 1992; Barry 1993). In marine systems, GPS has been used with dye tracer studies to monitor circulation patterns and water movement from sewage outfalls (Humphreys 1992).

Once a sample site has been identified, a GPS receiver can be used to navigate back to that site. Some types of GPS receivers are designed specifically for marine navigation, providing information on bearing and distance to the target. GPS units costing ~$100 are now commonly sold in fishing supply shops, and are used by fishermen to navigate to "honey holes," areas of lake bottom habitat in which fish congregate (Knowlton 1992).

Environmental quality — GPS can document the location of environmental hazards and treasures. Minnesota researchers are using GPS to locate groundwater wells sampled under the statewide Ground Water Monitoring and Assessment Program (Hsu *et al.* 1993). Biologists working for the Pennsylvania Department of Transportation mapped wetlands and other natural features to be avoided within a 50-mile, 1000-foot-wide corridor proposed for construction of a new state highway, using GPS to document their location (Halsch 1992).

Archeology and paleontology — GPS has been used to document the location of archeological and paleontological sites in remote areas (Miller 1991; Druss 1992; McKenna 1992). Once located, archeological sites can be compared with GIS data to determine how factors such as hydrology and topography influenced site selection for human habitation.

8.4 GPS and remote sensing

Imagery acquired by remote sensing can be converted into a GIS-usable format if georeferenced (see Chapter 9). Georeferencing is normally done by finding identifiable, discrete points on an image (e.g. road intersections, fence lines, buildings), determining their ground coordinates from a map or digital database, and matching the image to those coordinates. However, this method cannot be used in unmapped terrain or areas in which mapping is less precise than the resolution of the imagery. GPS can provide precise coordinates for ground control points in such instances, and has been used to georeference satellite imagery and aerial photography (Zavala & Garcia 1992; Bertram & Cook 1993).

Classification of a satellite image often involves computer "training" by specifying areas of known cover. This can pose a problem in extensive unmapped forests because of the difficulty of identifying ground locations on a relatively featureless image. GPS helped solve this problem in the rain forests of northern Guatemala, where field data on species composition, tree heights, and basal area per hectare were measured in plots georeferenced by GPS (Miller 1991). A light plane was chartered to photograph these areas with color-infrared film that produced a spectral signature similar to that of the satellite imagery, using the GPS coordinates to navigate to the proper areas. This low-altitude photography enabled interpreters to identify these areas on the satellite imagery and correctly classify forest cover types.

Low-altitude, low-cost means of remotely sensing ecological features, such as 35-mm cameras and airborne videography, are increasingly being used for ecological and forestry applications. Like other image types, however, low-altitude imagery must be georeferenced before use in a GIS. The lack of identifiable ground control points is particularly a problem when missions are flown at low altitudes, decreasing the likelihood that identifiable ground control points will be visible within an image frame. Use of a GPS in conjunction with these aircraft-borne devices can solve this problem. A GPS in an airplane records location data while flying over a target area. Although post-processing differential correction can be performed on the GPS data acquired, real-time differential is preferable, with a base station GPS positioned over a known location radioing correction data to the GPS in the airplane (Fig. 8.7).

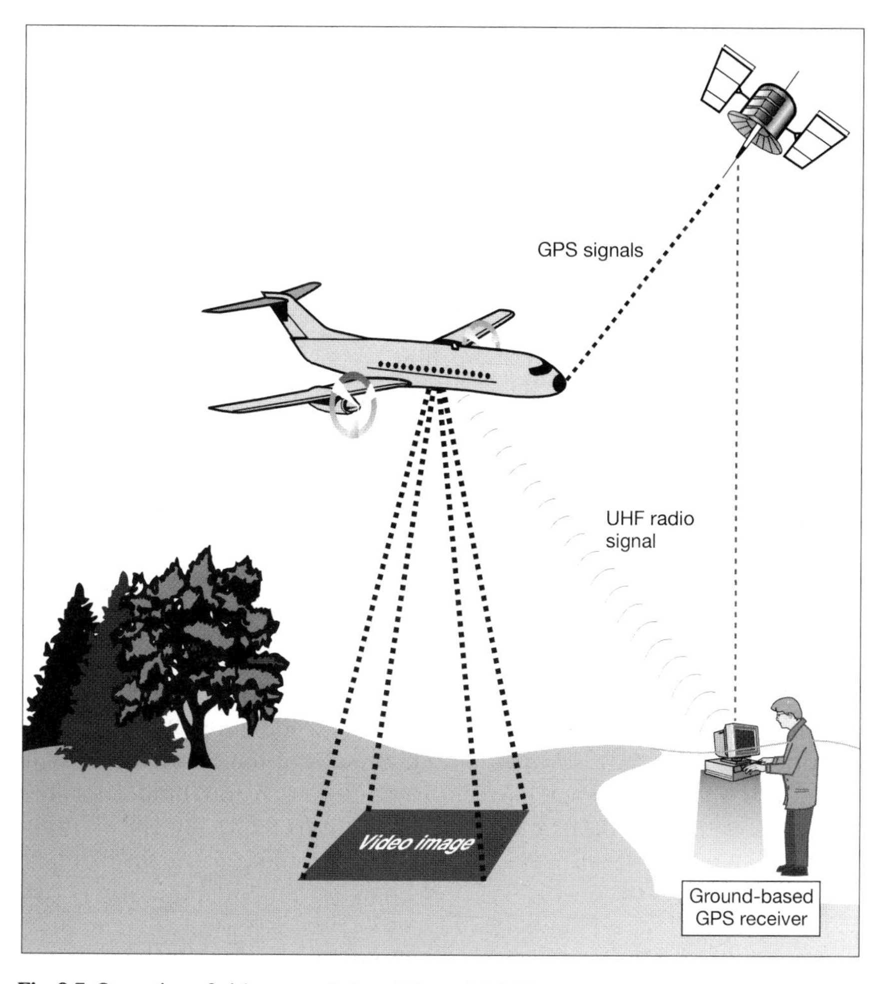

Fig. 8.7 Operation of airborne real-time differential GPS.

The GPS readings acquired during an aerial mission give the coordinates of the point directly beneath the airplane (its **nadir**), which is usually assumed to be the center of the image. If the scale and compass orientation of the image are known, the entire scene can then be georeferenced. However, aircraft tip and tilt can cause the center of the image to be offset from the nadir, and significant errors can result. GPS can solve this problem as well, through the use of aircraft attitude detection. Four GPS antennas are mounted to the front, back, and two sides of the fuselage, and collect data about the aircraft's orientation. This information can then be used to correct the georeferencing of the image.

There are several ways to record GPS readings in conjunction with airborne video. The locational coordinates of the nadir can be dubbed on the video image as a title caption (Bobbe 1992) or recorded on the audio track of the video tape. Alternatively, the GPS readings and SMPTE codes (a motion picture industry standard used to time-stamp video images) can be simultaneously recorded in a laptop computer attached to both the video camera and the GPS receiver (Sersland *et al.* 1995). To convert a video image to GIS-compatible format, a frame-grabber is used to select an individual video frame, and its nadir position is used to georeference the other pixels in the image. The resulting product is then ready for image analysis, discussed in Chapter 9.

CHAPTER 9

Remote sensing and GIS

Remote sensing is an important source of GIS data, because much of the information needed by ecologists is not already mapped (Ehlers *et al.* 1989; Davis *et al.* 1991; Trotter 1991; Vande Castle 1991). Remote sensing is used in the broad sense here to mean the acquisition of information about an entity without being in physical contact with it, including photography and videography as well as other imaging systems. Most remote sensing is done from airplanes and satellites, but unconventional platforms such as cherry pickers, helium balloons, and large model airplanes may also be useful in ecological applications (Hinckley & Walker 1993).

All remote sensing relies on the detection of eletromagnetic energy. Humans constantly perform remote sensing as they detect light energy to see, and sound waves to hear. The primary difference between looking across a room and looking down from space is a matter of scale, perspective, and experience. We are accustomed to the shape, size, and color of the table and chair across the room, and can therefore recognize and classify these objects. Most of us, however, cannot associate the shapes, sizes, and colors on a remotely sensed image with landscape features. This is partly because we can no longer distinguish objects that are familiar to us at the scale of individual organisms — we must learn to substitute the forest for the trees. This is often difficult in ecology, because much of our scientific experience is based upon individual organisms, rather than assemblages thereof.

This chapter introduces the basic concepts needed to apply remote sensing in a GIS context, but is by no means comprehensive. The reader who wishes to pursue these topics in more detail is referred to other sources of information on remote sensing (Colwell 1983; Jensen 1986; Lillesand & Kiefer 1994; Philipson 1996).

9.1 The electromagnetic spectrum

All remote sensing methods detect electromagnetic energy, which includes such familiar forms as visible light, X-rays, ultraviolet rays, television waves, and radio waves. The **wavelength** of different forms of electromagnetic energy is measured in units of micrometers (μm) to meters. The electromagnetic spectrum ranges from very short (10^{-7} μm) cosmic rays to the very long (> 10^{8} μm) waves used in radio and television transmission.

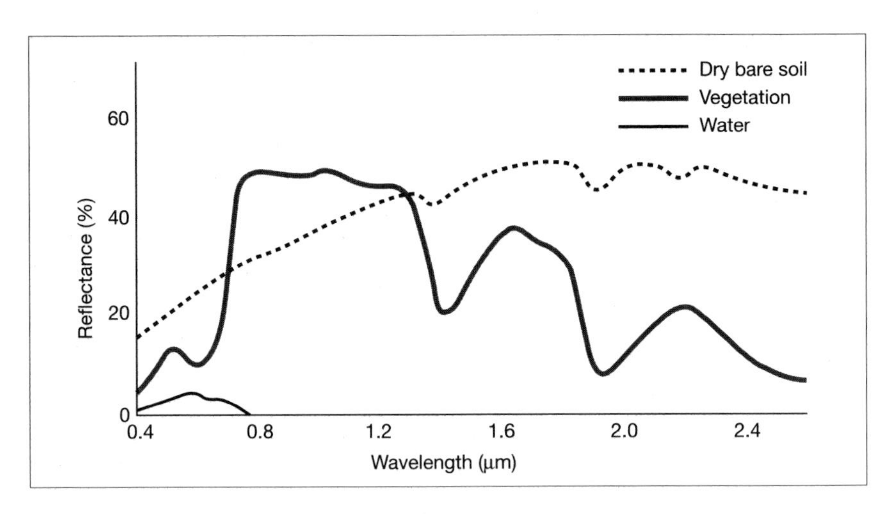

Fig. 9.1 Typical spectral reflectance curves for vegetation, soil, and water. From *Remote Sensing and Image Interpretation* by Lillesand & Kiefer, © 1994. Reprinted by permission of John Wiley & Sons, Inc.

The sun is the primary source of electromagnetic radiation to the Earth, particularly in the wavelength range of 0.3–10 μm. The **incident energy** that passes through the Earth's atmosphere interacts with the Earth's surface, where it may be **reflected**, **absorbed**, or **transmitted**. The reflected energy is detected by remote sensors, and differs for different materials and wavelengths. A graph of wavelength (x axis) versus per cent reflectance (y axis) is called a **spectral reflectance curve** (Fig. 9.1).

In addition to reflected solar energy, all matter at temperatures above absolute zero continuously **emit** electromagnetic radiation, though at a considerably different magnitude and spectral composition than that radiated by the sun (Lillesand & Kiefer 1994). The length of the dominant wavelength emitted is inversely proportional to the object's temperature, such that the maximum wavelength emitted by the Earth is about 9.7 μm, as opposed to about 0.5 μm for the sun.

Most remote sensors detect energy in one or several of the visible, infrared, and microwave portions of the electromagnetic spectrum, described below.

9.2 Energy interactions with Earth features

Light that is **visible** to the human eye is a very small portion of the electromagnetic spectrum. Blue light has wavelengths of approximately 0.4–0.5 μm, green light has wavelengths of approximately 0.5–0.6 μm, and red light has wavelengths of approximately 0.6–0.7 μm. The color of an object illuminated by visible light results from its reflectance in one or more of these primary color bands. We perceive healthy vegetation as being green because chlorophyll strongly absorbs blue and red light (wavelengths centered at 0.45 and 0.65 μm, respectively) and reflects green light.

Remote sensing allows humans to extend the limits of their vision to a broader spectrum of electromagnetic energy. Ultraviolet (**UV**) rays have wavelengths in the 0.3–0.4 μm range, but are absorbed or scattered as they pass through the atmosphere, and consequently are rarely used in remote sensing. Near-infrared (**NIR**) energy has wavelengths between 0.7 and 2.7 μm. Water, whether contained in water bodies, soils, or vegetation, strongly absorbs NIR energy, making water bodies appear dark on infrared photography. Vegetation, on the other hand, reflects much more NIR than visible light (Fig. 9.1). Reflectance of wavelengths in the 0.7–1.3 μm range from vegetation is due primarily to the internal structure of plant leaves (Lillesand & Kiefer 1994). Because this structure is highly variable among plant species and is affected by plant vigor, reflectance measurements in this wavelength range are very important to mapping vegetation and detecting plant stress. Bare soil reflects strongly in the NIR, particularly in the 1.5–2.6 μm range (Fig. 9.1). Dips in reflectance of NIR energy from vegetation and soil occur at 1.4, 1.9, and 2.7 μm because water contained within them absorbs strongly at these wavelengths (Lillesand & Kiefer 1994).

Wavelengths in the far-infrared or thermal-infrared (**TIR**) portion of the spectrum are used for detecting temperature differences. Peak energy emissions for most Earth features are in the 8–14 μm range, but peak emissivity of areas heated above the Earth's ambient temperature occurs at shorter wavelengths (3–5 μm). This characteristic is useful for monitoring forest fires and lava flows. TIR is also useful for making maps of surface temperature distributions, such as plumes of heated wastewater discharged into rivers.

The **microwave** portion of the electromagnetic spectrum includes wavelengths ranging from 1 mm to 2 m. Unlike visible light, microwaves can penetrate through clouds and atmosphere haze, which makes them very useful for remote sensing in tropical regions. Also, the factors that affect the reflection or emission of microwaves from Earth features are quite different from the factors that affect reflection of visible or thermal energy.

9.3 Sensors

Remote sensing systems are of two general types: **passive** and **active**. Passive remote sensing devices are most common, and use naturally available energy sources to detect Earth features. Examples are photographic cameras, video cameras, multispectral scanners, and thermal scanners. Active remote sensing systems, such as radar, supply their own source of energy to illuminate features of interest.

9.3.1 Passive systems

Photographic cameras use film as the medium to detect reflected energy. Color and panchromatic black and white films are sensitive to light in the UV and visible range, but the UV (0.3–0.4 μm) is usually blocked with a "haze" filter. Infrared films detect energy from the UV to NIR portions of the spectrum (0.3–0.9 μm), but wavelengths shorter than 0.5 μm are usually blocked with a "blue-absorbing" filter. In color and color-infrared films, the different portions of the electromagnetic spectrum expose different layers of yellow, magenta, and cyan dye, which combine to make red, green, and blue. Healthy vegetation reflects NIR energy more strongly than other wavelengths (Fig. 9.1), which causes the developed film to appear red (= yellow + magenta). Color-infrared film is often referred to as "false-color" because the colors depicted are different from those which we see in nature.

Digital cameras have similar optics as do photographic cameras, but use electronics instead of film to record light intensity. Similar to the data structure used by raster GIS, each image is divided into thousands of tiny pixels. Digital cameras employ arrays of charge couple devices (**CCD**s), each of which senses one pixel in the image field. Array sizes for digital cameras typically range from 512 × 512 pixels to 2048 × 2048 pixels or more. The output of digital cameras is of lower resolution than a photograph, but has the advantage of being produced in near real-time, not requiring film development. Also, as the name implies, the output from digital cameras is already in digital format, and can be imported directly into an image analysis or GIS program. Color, color-infrared, and **multispectral** digital cameras are available. Multispectral systems detect several **bands** of spectral energy (e.g. blue, green, red, NIR).

Video cameras also detect energy in the visible to NIR range, but instead of recording to film, they record to magnetic tape. In consumer-grade video cameras, incoming light is diffracted by a prism to separate the blue, green, and red wavelengths, which are converted to an analog signal based on scene brightness. Instead of a prism, professional-grade video cameras use separate CCDs for different wavelength bands, Multispectral video cameras are also available, and have demonstrated utility for vegetation mapping (Nixon *et al.* 1985; Mausel *et al.* 1989; Bartz *et al.* 1992; Bobbe *et al.* 1994; Thomasson *et al.* 1994). To extract information from the video tape, a frame-grabber is used to select individual images, rasterize the images, and convert them to a GIS-compatible format. Like digital cameras, videotaped images are immediately available for viewing and do not require film development. The principal disadvantage of video cameras is their poor spatial resolution relative to film or digital cameras.

A **scanner** builds a two-dimensional image from scan lines using detectors that produce electrical signals proportional to the energy received from Earth surfaces. The forward motion of the aircraft causes a new strip of ground to be covered by successive scan lines (Fig. 9.2). **Electro-optical** scanners contain a mirror that rotates from side to side along scan lines oriented perpendicular to the path of the aircraft,

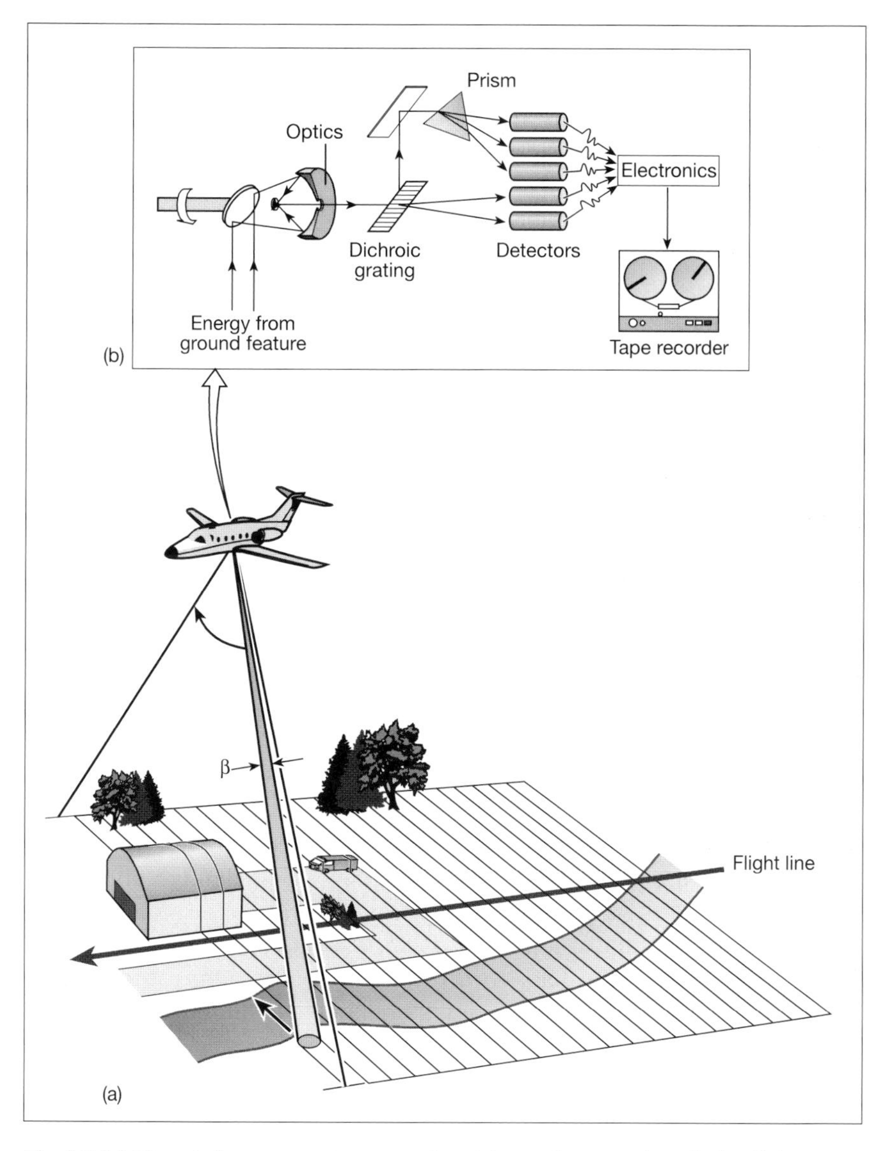

Fig. 9.2 Multispectral scanner system operation: (a) scanning procedure during flight, (b) scanner-schematic. From *Remote Sensing and Image Interpretation* by Lillesand & Kiefer, © 1994. Reprinted by permission of John Wiley & Sons, Inc.

whereas **pushbroom** scanners contain a one-dimensional linear array of detectors that sense the scan line simultaneously. The SPOT satellite (see Section 9.4.3) sensors use pushbroom scanners, whereas the older Landsat system (see Section 9.4.1) uses a moving mirror design. There are scanners designed for a variety of wavelengths, including visible, NIR, TIR, and microwaves. Satellite remote sensing is done almost exclusively with scanners.

9.3.2 Active systems

Radar is an active remote sensing system. Radar transmits short bursts of microwave energy in the direction of interest and records the strength and origin of "echoes" or "reflections" received from objects within the system's field of view (Lillesand & Kiefer 1994). Because active systems do not rely on reflected light, imagery can be acquired day or night.

Doppler radar systems utilize Doppler Effect frequency shifts in the transmitted and returned microwave signals to determine an object's velocity. This remote sensing capability is commonly used by law enforcement officers to monitor vehicle speeds. Of more utility for GIS applications is the use of Doppler radar to monitor the spatial distribution of precipitation. The U.S. National Oceanic and Atmospheric Administration (NOAA) has constructed a network of NEXRAD Doppler radar sites which provide digital images of precipitation amounts commonly used in television weather programs. These precipitation data are spatially distributed, making them much more accurate than data interpolated from weather monitoring stations (Klazura & Imy 1993).

Airborne radar remote sensing, called side-looking airborne radar (**SLAR**), sends and receives radar pulses from an antenna fixed below the aircraft and pointed to the side, producing continuous strips of imagery covering a swath of ground adjacent to the path of the airplane (Fig. 9.3). This antenna arrangement is used because for any given wavelength, the larger the antenna, the better the spatial resolution (Lillesand & Kiefer 1994). As it uses microwaves, SLAR can penetrate cloud cover, and has been instrumental in mapping such perennially cloudy areas as the rainforests of Panama and the Amazon Basin.

Synthetic aperture radar (**SAR**) systems employ a short physical antenna, but synthesize the effect of a long antenna through modified data recording and processing techniques (Lillesand & Kiefer 1994). RADARSAT, Canada's entry into the arena of Earth observation satellites, began acquiring imagery in 1995 with a SAR sensor (Nazarenko *et al.* 1996).

Acoustic remote sensing devices are used for the underwater detection of fish. Acoustic systems sample the water column by sending short (e.g. 0.2–1.0 m/s), repetitive (e.g. five per second) pulses of high-frequency sound (e.g. 12–420 kHz) in a beam downward through the water column

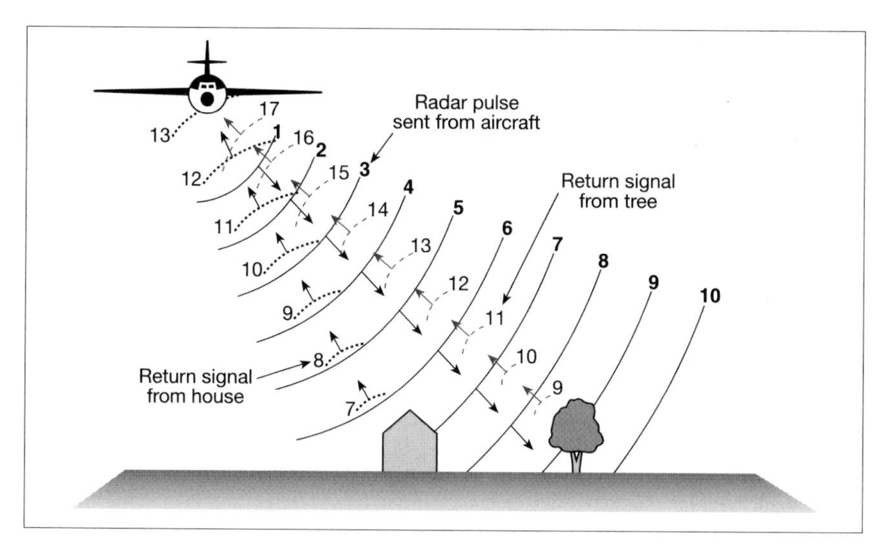

Fig. 9.3 Operation of a side-looking airborne radar. From *Remote Sensing and Image Interpretation* by Lillesand & Kiefer, © 1994. Reprinted by permission of John Wiley & Sons, Inc.

(Brandt *et al.* 1996). A two-dimensional vertical transect is recorded as the survey vessel moves across the water surface. When the sound wave encounters an acoustic scatter, i.e. a fish, an echo returns to the ship, providing information about fish location, size, numerical abundance, and biomass. Acoustic data are converted into cells by defining the depth intervals and horizontal distances over which the continuous data are pooled, providing a vertical continuous data layer of fish densities and sizes (Brandt *et al.* 1996).

9.4 Satellite imagery

A number of countries have launched satellites for image acquisition, some of which are listed in Table 9.1. Satellite imagery has several advantages as a source of GIS data:

- already in digital form;
- frequent recurrence of coverage;
- a single image covers an extensive area, greatly reducing the time required for rectifying, georeferencing, and edge-matching multiple images;
- image analysis can provide quantitative information about ecological properties which cannot be easily derived from aerial photography or field studies (e.g. NDVI – see Section 6.2).

The major disadvantage of satellite imagery as a source of GIS data is its relatively poor spatial resolution. Satellite images can provide data about plant communities and environmental conditions, but are unsuit-

Table 9.1 Some operational and research satellites for Earth surface analysis. After Davis and Simonett, 1991.

Platform	Sensor	Year	No. of bands	Spectral coverage	IFOV	Repeat	Country	Scene width (km)
Landsat	MSS	1972–	4	VIS/NIR	80 m	16 days	U.S.	185
	TM	1982–	7	VIS/NIR/TIR	30/30/120 m	16 days	U.S.	185
NOAA	AVHRR	1978–	5	VIS/NIR/TIR	1–4 km	12 hours	U.S.	2400
Shuttle	SIR-A	1981	1	Radar	40 m	–	U.S.	50
	SIR-B	1984	1	Radar	14–46 m	–	U.S.	20–50
SPOT	HRV-P	1986–	1	VIS	10 m	26 days	France	117
	HRV-XS	1986–	3	VIS/NIR	20 m	26 days	France	117
ERS-1	AMI	1991	1	Radar	30 m	3 days	E.C.	100
	ASTR	1991	3	TIR	1 km	3 days	E.C.	100
JERS-1	OPS	1992	7	VIS/NIR	18 m	44 days	Japan	75
	SAR	1992	1	Radar	18 m	44 days	Japan	75
RADARSAT	SAR	1995	1	Radar	25 m	3 days	Canada	45–500

able for studies of individual organisms. Clouds and atmospheric effects also interfere with satellite imagery, because the reflected energy must pass through the Earth's atmosphere before reaching the satellite in space. Furthermore, because satellite imagery relies on spectral reflectance, it is applicable only to features and phenomena that have distinguishable spectral characteristics.

9.4.1 Landsat

Landsat was the first non-military satellite image acquisition program (Table 9.1). Landsat satellites -1, -2, and -3 were launched by the U.S. in 1972, 1975, and 1978, respectively, and each was decommissioned about 5 years after launch. These early Landsat satellites carried Multi-Spectral Scanner (**MSS**) sensors, which detect wavelengths in four spectral bands: green (0.5–0.6 μm), red (0.6–0.7 μm), and two bands of NIR (0.7–0.8 and 0.8–1.1 μm). These band widths are large by today's standards, as is the 79-m instantaneous field of view (**IFOV**), giving MSS imagery rather poor spectral and spatial resolution. Nevertheless, MSS data are still useful for change detection and multitemporal image analysis (Wolter *et al.* 1995).

Subsequent Landsat satellites carried Thematic Mapper (**TM**) scanners in addition to MSS scanners. TM has seven spectral bands, including three in the visible spectrum, one NIR band that is narrower than the NIR band of MSS, two mid-infrared bands, and one TIR band. These additional bands provide TM imagery with better vegetation discrimination than that of MSS (Table 9.2). TM imagery also has much finer spatial resolution than its Landsat predecessors, which is desirable for ecological applications: the IFOV is 30 m for all bands except the TIR, which has a 120-m IFOV. TM has been used for mapping a variety of ecological features: national land cover (Scott *et al.* 1993; Fuller *et al.* 1994), river flooding (Michener & Houhoulis 1996), grassland vegetation (McGuire *et al.* 1990; Lauver & Whistler 1992), tundra (Joria & Jorgenson 1996), forests (Congalton 1993; Wolter *et al.* 1995), deforestation (Tucker *et al.* 1984), marine habitats (Luczkovich *et al.* 1993; Zainal *et al.* 1993), and wildlife habitats (Ormsby & Lunetta 1987; Hodgson 1988; Griffiths *et al.* 1993; Herr & Queen 1993). Landsat satellites have sun-

Table 9.2 Landsat Thematic Mapper spectral bands. *Remote Sensing and Image Interpretation* by Lillesand & Kiefer, © 1994. Reprinted by permission of John Wiley & Sons, Inc.

Band	Wavelength (μm)	Nominal spectral location	Principal applications
1	0.45–0.52	Blue	Designed for water body penetration, making it useful for coastal water mapping. Also useful for soil/vegetation discrimination, forest type mapping, and cultural feature identification.
2	0.52–0.60	Green	Designed to measure green reflectance peak of vegetation (Fig. 9.1) for vegetation discrimination and vigor assessment. Also useful for cultural feature identification.
3	0.63–0.69	Red	Designed to sense in a chlorophyll absorption region (Fig. 9.1) aiding in plant species differentiation. Also useful for cultural feature identification.
4	0.76–0.90	NIR	Useful for determining vegetation types, vigor, and biomass content, for delineating water bodies, and for soil moisture discrimination.
5	1.55–1.75	Mid-infrared	Indicative of vegetation moisture content and soil moisture. Also useful for differentiation of snow from clouds.
6*	10.4–12.5	TIR	Useful in vegetation stress analysis, soil moisture discrimination, and thermal mapping applications.
7*	2.08–2.35	Mid-infrared	Useful for discrimination of mineral and rock types. Also sensitive to vegetation moisture content.

*Bands 6 and 7 are out of wavelength sequence because band 7 was added to the TM late in the original design process.

synchronous orbits with a 16-day repeat cycle for each satellite, but the orbits of Landsat-4 and Landsat-5 were established 8 days out of phase, such that an 8-day cycle could be maintained with alternating coverage by each satellite (Lillesand & Kiefer 1994).

9.4.2 Advanced Very High Resolution Radiometer (AVHRR)

Meteorological satellites launched by the U.S. NOAA contain the Advanced Very High Resolution Radiometer (**AVHRR**). AVHRR has an IFOV of 1.1 km at nadir, a spatial resolution which is quite coarse for most ecological applications. All of the data are resampled at a nominal resolution of 4 km, with only selected data recorded at the full 1.1 km resolution. NOAA satellites provide twice-daily coverage of the Earth, the even-numbered missions crossing the equator in the western hemisphere during daylight hours and the odd-number missions crossing at night. The AVHRR detects red light (0.58–0.68 μm), NIR (0.72–1.10 μm), and two to three bands of TIR, depending on the satellite. Its 2400-km swath width covers large regions simultaneously (Fig. 9.4).

AVHRR's frequent repeat rate and thermal detection capabilities are useful in studies of ephemeral phenomena such as wildfire (Riggan *et al.* 1993; Chuvienco & Martín 1994), water temperature (Leshkevich *et al.* 1993), and lake ice (Wynne & Lillesand 1993). AVHRR has also been used for large-area forest mapping and land cover characterization (Brown *et al.* 1993; Ripple 1994; Stone *et al.* 1994; Zhu & Evans 1994). In the U.S., AVHRR is the source imagery for the multitemporal Land Cover Characterization Program (Van Driel & Loveland 1996) described in Section 6.3.

Fig. 9.4 AVHRR image of Lake Superior in North America (0.72–1.10 μm band), showing continuous ice cover on February 9, 1994. Courtesy of Peter Wolter, Natural Resources Research Institute, University of Minnesota, Duluth.

9.4.3 Systeme Pour l'Observation de la Terre (SPOT)

In early 1986 the French government, in cooperation with Sweden and Belgium, launched the Systeme Pour l'Observation de la Terre (**SPOT**). SPOT employs pushbroom scanners that detect green (0.50–0.59 μm), red (0.61–0.68 μm), and NIR bands (0.79–0.89 μm) with a 20-m IFOV, and a panchromatic (0.51–0.73 μm) sensor with a 10-m IFOV. The orbit pattern for SPOT repeats every 26 days, but because SPOT has optics that can be pointed, an area of interest can be imaged by satellite passes that do not pass directly overhead. Stereoscopic imaging is possible by recording overlapping swaths on successive days. In addition to its applicability for mapping vegetation (McGuire *et al.* 1990; Rutchey & Vilcheck 1994; Welch *et al.* 1995) and wildlife habitat (Miller & Conroy 1990), SPOT has also been used to characterize landscape structure in Belgium and Zaire (Goossens *et al.* 1993; Gulinck *et al.* 1993).

9.4.4 Radar satellites

Several satellites carrying radar sensors were launched in the early 1990s, including the European Space Agency's ERS-1 satellite, the National Space Development Agency of Japan's JERS-1, and the Canadian RADARSAT (Table 9.1). There are fewer examples of the application of satellite radar imagery in ecology, but the Shuttle Imaging Radar (SIR) experiments conducted from the U.S. space shuttle in the 1980s showed promise for forest-related applications (Imhoff & Gesch 1990; Leckie 1990).

9.5 Extracting information from remotely sensed data

9.5.1 Aerial photographs

Aerial photography came into widespread use in the late 1930s, and is still an essential source of information for a variety of purposes. Aerial photography is suitable for long-term change detection because of its routine acquisition over multiple decades by many government agencies (see Chapter 6).

Aerial photography comes in a variety of formats. Large-format cameras designed specifically for aerial photography expose a 230 × 230 mm negative, but vertically mounted 35-mm cameras can provide satisfactory results for covering small areas (Befort 1986; Gilruth & Hutchinson 1990). The earliest aerial photographs used panchromatic black and white film, which has been used in a variety of natural resource applications, particularly those involving analysis of change over time (Johnston & Naiman 1990; Welch & Remillard 1992; Green *et al.* 1993; Johnson & Johnston 1995). Color film is preferred for interpretation of underwater features, because of its superior water penetration (Ferguson *et al.* 1993).

Black and white infrared film has been used in forestry applications because deciduous and coniferous trees are more easily distinguished than with panchromatic film (Zsilinszky 1966; USDA 1978), but color-infrared film is more commonly used in vegetation studies (Welch *et al.* 1988; Pastor & Broschart 1990; Tiner 1990; Bobbe *et al.* 1992).

The scale of an aerial photograph can be determined as follows:

1 Find two points on the aerial photograph that are at about the same elevation and a known distance apart on the ground (one way to determine this is to measure the distance between them on a map, and use the map scale to figure their ground distance). Avoid points that are at the edges of the photograph.

2 Use an engineer's scale or another finely divided measuring device to measure the distance between the two points on the photograph.

3 Convert the photograph distance and the ground distance to common units of measure.

4 Compute the photograph scale as a ratio of the photograph distance to the ground distance, and express it as a representative fraction (see Section 2.2.6). For example, if the ground distance between two points is equal to 1.61 km, and the measured distance between those two points on the aerial photograph is 4.03 cm, then the representative fraction is:

$$RF = \frac{\text{photograph distance}}{\text{ground distance}} = \frac{4.03 \times 10^{-2}\ \text{m}}{1.61 \times 10^{3}\ \text{m}} = \frac{1}{39\,950} \qquad (9.1)$$

Aerial photograph scale is also inversely proportional to the altitude of the aircraft relative to the terrain being photographed:

$$RF = \frac{\text{camera focal length}}{\text{flying height above terrain}} \qquad (9.2)$$

Therefore, a high-altitude (small-scale) photograph covers a larger area than a low-altitude (large-scale) photograph taken with the same focal length lens.

The information content of an aerial photograph must be interpreted to be made meaningful. The human brain can process considerable information about an aerial photograph, and use that information to recognize and delineate landscape entities. Seven basic characteristics of photographic features are used in this process (Lillesand & Kiefer 1994):

- **tone** (black and white photographs) or **color**;
- **shape**;
- **size**;
- **pattern** — the spatial arrangement of objects;
- **texture** — the frequency of tonal change in the photographic image;

- **shadows**;
- **site**.

A trained aerial photograph interpreter uses these clues in combination with his or her knowledge of landscape features to make an identification. For example, a light-toned, uniform textured circular patch 0.8 km in diameter located in the U.S. Great Plains is probably a crop of small grains irrigated by pivot irrigation. Sometimes these clues are clear-cut, as in the case of pivot irrigation, and sometimes they are more subtle.

Stereoscopic aerial photograph interpretation provides the additional information of the third dimension. The height of tall objects such as buildings can be observed and measured, and topography becomes evident. Stereoscopic air photograph interpretation is important in forest mapping because of the clues provided by different tree heights and crown shapes (Zsilinszky 1966; USDA 1978), and in wetland and soil mapping because of the clues provided by landscape position (Soil Survey Staff 1966; Johnston *et al.* 1988; Tiner 1990).

Aerial photographs are often used in mapping land use and land cover. **Land cover** is the material covering a land surface, such as vegetation, exposed soil, bedrock, etc., whereas **land use** is the human-imposed function of a land area (Star & Estes 1990). Distinguishing land use requires the knowledge of human behavior that only another human can provide. Therefore, aerial photograph interpretation has been used to prepare land use maps for a number of nations (Fegeas *et al.* 1983; Aspinall & Pearson 1996).

A hierarchical land use/land cover classification developed by Anderson and co-workers (1976) for use with aerial photography and remotely sensed data is still widely used. The coarsest level of classification consists of nine categories (Level I: Table 9.3), and is generically referred to as "Anderson Level I." Each of those categories is further subdivided into Level II categories, a classification resolution suitable for use on a nationwide basis. Levels III and IV provide additional detail, to be designed by local users.

An alternative to conventional aerial photograph interpretation is to digitize aerial photographs using an image scanner, and apply digital image classification techniques (see below) to the digital data (Scarpace *et al.* 1981).

Aerial photographs are **unrectified**, meaning that they contain image displacement due to elevation differences, aircraft tip and tilt, and lens distortion (Bolstad 1991). Such displacement impairs accurate point location and measurement of distances, areas, or angles directly off a photograph. Although it is technically possible to digitize landscape boundaries directly from an aerial photograph, photograph displacement

Table 9.3 USGS land use and land cover classification system for use with remote sensor data. From Anderson *et al.* 1976.

Level I	Level II
1 Urban or built-up land	11 Residential
	12 Commercial and services
	13 Industrial
	14 Transportation, communications, and utilities
	15 Industrial and commercial complexes
	16 Mixed urban or built-up land
	17 Other urban or built-up land
2 Agriculture land	21 Cropland and pasture
	22 Orchards, groves, vineyards, nurseries, and ornamental horticulture areas
	23 Confined feeding operations
	24 Other agricultural land
3 Rangeland	31 Herbaceous rangeland
	32 Shrub and brush rangeland
	33 Mixed rangeland
4 Forest land	41 Deciduous forest land
	42 Evergreen forest land
	43 Mixed forest land
5 Water	51 Streams and canals
	52 Lakes
	53 Reservoirs
	54 Bays and estuaries
6 Wetland	61 Forested wetland
	62 Non-forested wetland
7 Barren land	71 Dry salt flats
	72 Beaches
	73 Sandy areas other than beaches
	74 Bare exposed rock
	75 Strip mines, quarries and gravel pits
	76 Transitional areas
	77 Mixed barren land
8 Tundra	81 Shrub and brush tundra
	82 Herbaceous tundra
	83 Bare ground tundra
	84 Wet tundra
	85 Mixed tundra
9 Perennial snow or ice	91 Perennial snowfields
	92 Glaciers

will distort the resulting GIS data layer. Displacement caused by elevation differences and aircraft tip and tilt is especially severe in high altitude photographs. Several methods are used to compensate for this distortion:

1 Transferring the information to rectified photograph. The transfer can be done by eyesight because the same landscape boundaries are generally visible on both photographs. This procedure has been used for wetland maps (Johnston *et al.* 1988) and soil maps produced by the U.S. National Cooperative Soil Survey.
2 Transferring the information to a base map using a Bausch and Lomb Zoom Transferscope (ZTS). A ZTS allows the operator to superimpose a monoscopic or stereoscopic aerial photograph image on a base map, simultaneously viewing both. A zoom lens can be used to enlarge the scale of the aerial photograph to the scale of the base map, and an anamorphic lens system can be used to stretch the photograph image up to twofold larger in one direction.
3 Digitizing boundaries directly from the aerial photograph with a digitizing table and puck (see Section 1.4.1), and geometric transformation as described in Section 3.3.1.2. If the base data layer used to provide the control points for the transformation is georeferenced, this method will provide georeferencing as well as rectification.
4 Using a digital scanner to digitize the aerial photograph, "rubber sheeting" (see Section 3.3.1.2) the image, and performing "heads-up" digitizing to delineate features. Heads-up digitizing is a monoscopic interpretation method performed with a mouse (or another pointing device) while the operator is viewing the photographic image on the computer monitor.

In addition to being unrectified, aerial photographs are also not georeferenced (see Section 2.4). Methods 3 and 4 above can provide georeferencing if the base maps and control points for geometric transformation are georeferenced.

Orthophotographs are corrected photographs in which displacement errors have been removed by a process called differential rectification. Orthophotographs look like traditional aerial photographs but show the true position of objects and can be used like conventional maps. To create computer-readable orthophotographs, aerial photographs are scanned and then adjusted using aerotriangulation control points and terrain information (see Section 4.1.1). The resulting digital orthophotograph image can be displayed on a computer monitor or sent to a plotter to produce a paper or film graphic (Fig. 9.5). Digital orthophotographs are especially suitable for heads-up digitizing, because they are already rectified.

Several nations have begun to generate digital orthophotography as a basic GIS resource (Gunnarsson 1993; Light 1993). The U.S. Geological Survey (USGS) is developing a national program to create digital orthophotographs from existing National Aerial Photography Program (NAPP) products: digital orthophoto quads (**DOQ**) which correspond to

Fig. 9.5 A GIS road layer superimposed on a portion of the USGS DOQ for Duluth, Minnesota. Courtesy of Gerald Sjerven and Jim Salés, Natural Resources Research Institute, University of Minnesota, Duluth.

the area of individual USGS 7.5-minute quadrangles, and digital orthophotographs created from NAPP quarter-quad centered photography (digital orthophoto quarter-quads; **DQQ**). Each DQQ file represents a black and white NAPP 1 : 40 000 photographs scanned at a raster size of 25 μm, resulting in a ground resolution of about 1 m, and a file size of about 55 Mb. The files are georeferenced, so that other GIS data layers can be superimposed on them (Fig. 9.5).

9.5.2 Airborne videography

Airborne videography is a low-cost means of rapidly acquiring large-scale imagery for a variety of ecological and non-ecological purposes. Airborne videography has the advantage that image acquisition can be timed to coincide with an event of interest (e.g. a flood, an algae bloom) at a spatial scale suitable for detecting objects as small as individual organisms (e.g. trees, geese). Video technology is commonly available and inexpensive: acquisition of video imagery for habitat studies of the Platte River costs \$275 a flight, compared to >\$9000 for medium-scale aerial photography of the same area (Sidle & Ziewitz 1990). The data recorded are already in digital form, facilitating their conversion into a GIS database.

Airborne videography has been used to detect crop damage and weed infestations (Manzer & Cooper 1982; Escobar *et al.* 1983; Richardson *et al.* 1985), assess soil salinity (Everitt *et al.* 1988), map pipeline corridors (Maggio *et al.* 1993), distinguish plant species and estimate phytomass production in grasslands (Nixon *et al.* 1985; Everitt & Nixon 1985; Everitt *et al.* 1986), map upland and wetland plant communities (Jennings *et al.* 1992; Bartz *et al.* 1992; Sersland *et al.* 1995), map forests (Bobbe *et al.* 1994; Thomasson *et al.* 1994), measure water turbidity (Mausel *et al.* 1989), and assess habitat for waterfowl and shorebirds (Cowardin *et al.* 1989; Sidle & Ziewitz 1990).

Like aerial photography, airborne videography is unrectified and not georeferenced. Geometric transformation techniques can be used to fit video images to known ground locations in populated or cultivated areas (e.g. road intersections, field corners), but ecologists generally work in terrain that lacks the anthropogenically derived rectilinear features needed for rubbersheeting. The lack of features for georeferencing is particularly a problem for ecologists researching large lakes or extensive wetlands. Methods for overcoming this deficiency often negate the advantages of using airborne video: (i) flying at higher elevations to include known ground locations within the field of view of an individual video frame decreases resolution and classification accuracy, and (ii) placement of field targets at known locations along the flight path is very labor-intensive, especially in roadless areas.

Global Positioning Systems (GPS — see Chapter 8) can be linked to video cameras to record aircraft position as imagery is being acquired (Sersland *et al.* 1995). A GPS reading taken at the instant of video image acquisition represents the nadir of the aircraft (i.e. the point directly below it), which is not necessarily the center point of the image. If the camera is in perfect vertical orientation relative to the ground, then the aircraft nadir is the center point of the image, and the only image distortion in flat terrain is radial distortion (i.e. slightly smaller scale at the edges of the image). However, if the video camera is not vertical because of aircraft tip or tilt (a common occurrence at the low altitudes normally flown during airborne video missions), then the GPS reading will not correspond with the central pixel in the image, and the scale will differ from one side of the image to the other (Fig. 9.6).

Two approaches can be used to overcome this georeferencing problem: (i) use of a stabilized camera mount for maintaining vertical camera orientation, and (ii) use of a GPS capable of precisely measuring aircraft altitude. The former solution is expensive: gyroscope-operated camera mounts that are commercially available to maintain vertical camera orientation for aerial photography and videography cost about $65 000. The latter solution is slightly less expensive: an Ashtech 3DF ADU GPS receiver system, consisting of four GPS antennas mounted to each wing of the aircraft and the front and back of the fuselage, costs about $20 000.

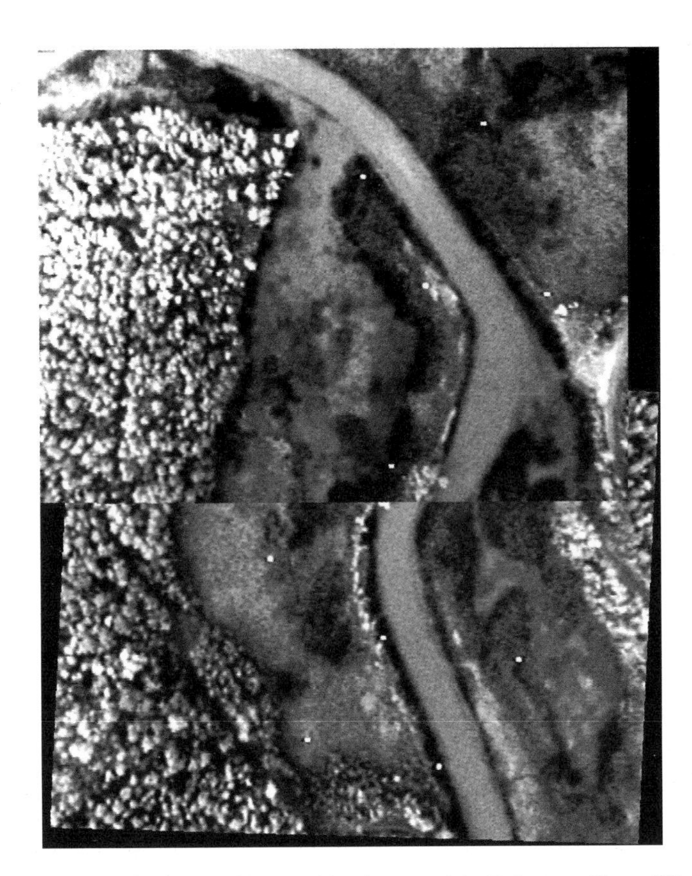

Fig. 9.6 Mismatch of adjacent airborne video images of the Pokegama River, Wisconsin due to georeferencing error induced by aircraft tip and tilt. Courtesy of John Bonde, Natural Resources Research Institute, University of Minnesota, Duluth.

Less expensive alternatives are currently being explored (Sersland *et al.* 1995; Johnston *et al.* 1996).

9.5.3 Digital image classification

In contrast to human interpretation techniques, in which information is derived from the visual inspection of images, digital image classification derives information from the numerical representation of an image. Digital classification is the conventional approach for analyzing scanner data, and is also used with digital aerial photographs and frame-grabbed video imagery.

Digital image classification methods can be divided into four categories (Lillesand & Kiefer 1994):

1 **Spatial pattern recognition** involves the classification of pixels based on their spatial relationship with surrounding cells. In the description given in Section 9.5.1 of the pivot-irrigated cropland, the characteristics

"uniform textured circular patch 0.8 km in diameter" are spatial patterns which could be detected automatically. Spatial classifiers consider such attributes as image texture, pixel proximity, directionality, repetition, context, and feature shape and size. Many of these types of analyses are basic to raster GIS, as discussed in Chapter 3.

2 **Temporal pattern recognition** uses temporal change as an aid to classification (see Section 6.3).

3 **Supervised spectral classification** uses the analyst's knowledge of ground features to aid the computer analysis of their spectral characteristics. As we have seen in Fig. 9.1, different landscape features have different spectral reflectance characteristics. The digital number (DN) values that represent these spectral properties are analyzed in spectral classification.

4 **Unsupervised spectral classification** uses mathematical algorithms to identify natural groupings of DN values, which are subsequently classified by the analyst. Supervised and unsupervised classification methods are described in more detail below.

9.5.3.1 Supervised classification

In supervised classification, areas known to have certain ground characteristics, called training sets, are delineated by the analyst on a displayed digital image. The distribution of digital numbers for the pixels in each training set is determined as illustrated by the scatterplot in Fig. 9.7. The computer develops numerical descriptions of these training sets, called "classifiers," and applies them to each pixel in the image. Four different classifiers are illustrated in Fig. 9.8.

A **minimum-distance-to-means-classifier** computes a centroid (mean data value) for the cluster of pixels in each training set, and classifies pixels outside of the data set based on their distance in spectral space from the nearest centroid (Fig. 9.8a). Pixels with distance-to-means values exceeding a threshold value are classified as "unknown." In Fig. 9.8a, pixel number 1 would be assigned to the corn class by this method, whereas pixel number 2 would be assigned to the sand, even though it falls on the edge of the cluster of "urban" pixels. This classifier is thus insensitive to the variance in the spectral data.

A **parallelepiped classifier** fits a rectangle around the range of DN values for the cluster of pixels in each training set. This method is very sensitive to the variance of the spectral data. In Fig. 9.8b, pixel number 2 would correctly be assigned to the urban class, but classification of pixel number 1 would be uncertain because of the overlap of the rectangles for corn and hay. This problem can be somewhat alleviated by placing a stepped rectangular boundary around the cluster of training cell pixels (Fig. 9.8c).

A **Gaussian maximum likelihood classifier** computes probability den-

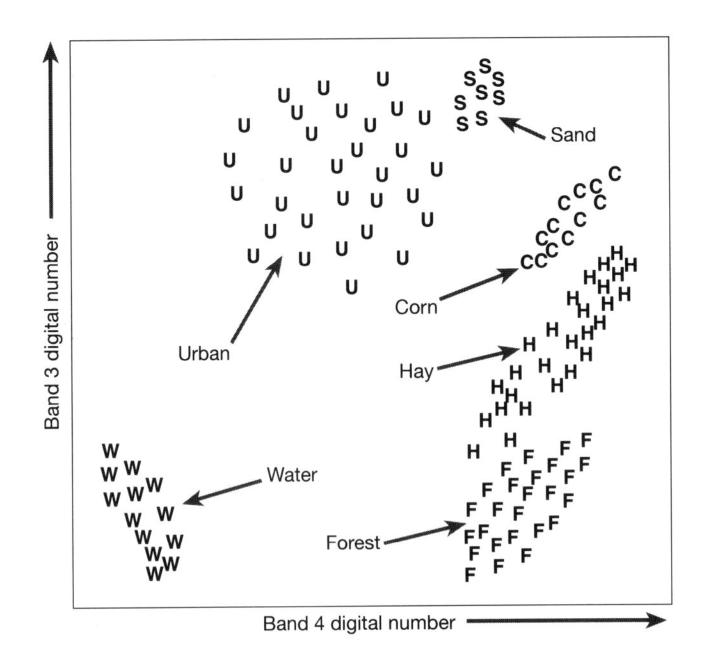

Fig. 9.7 Spectral scatterplot for supervised classification training sets. *Remote Sensing and Image Interpretation* by Lillesand and Kiefer, © 1994. Reprinted by permission of John Wiley & Sons, Inc.

sity functions for each cluster of training set pixels, essentially delineating ellipsoidal equiprobability contours for each class (Fig. 9.8d). The shape of the equiprobability contours is sensitive to covariance, so that pixel number 1 is now correctly classified as corn.

The **Bayesian classifier** is based upon another statistical approach, further discussed in Section 10.4.3.2. This technique applies two weighting factors to the probability estimate (Lillesand & Kiefer 1994). The *a priori* probability is a weight based on the likelihood of occurrence for each class, whereas the *a posteriori* probability is a weight representing the "cost" of misclassification. These two factors optimize the classification by minimizing the cost of misclassification.

Although the above classifiers were developed for use in remote sensing, they are equally applicable to classification of continuous data in a raster GIS. For example, instead of using digital numbers in different spectral bands as the parameter of measurement, we could use elevation and aspect data. Areas of known vegetation could be used to derive training set data from a digital elevation model (DEM – see Chapter 4). The elevation and aspect characteristics would be displayed as a scatterplot, and one or more of the above classifiers would be applied to those data. That classifier would be applied to the entire DEM to predict vegetation distribution based on elevation.

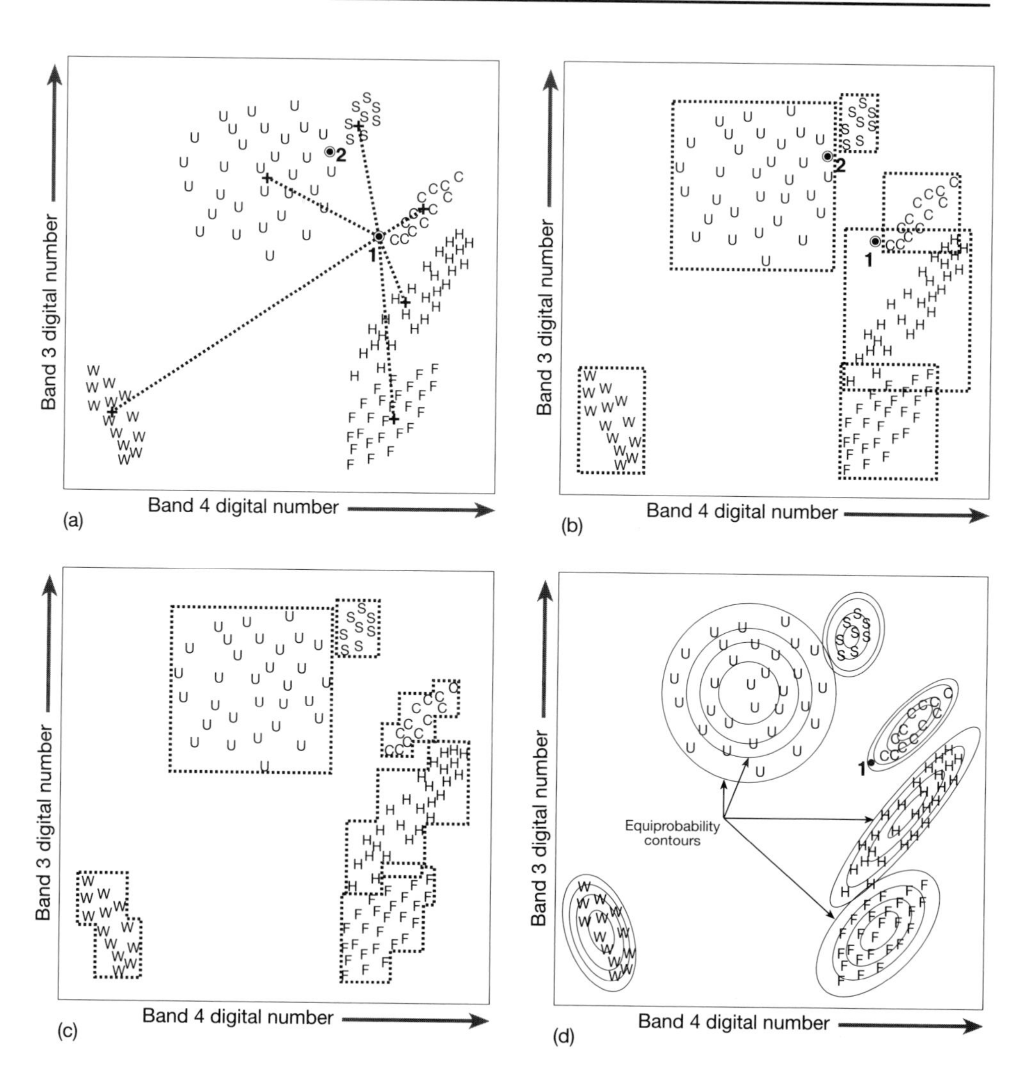

Fig. 9.8 Unsupervised classification algorithms: (a) minimum distance to means classification, (b) parallelepiped classification, (c) parallelepiped classification with stepped boundaries, (d) maximum likelihood classifier. *Remote Sensing and Image Interpretation* by Lillesand and Kiefer. Reprinted by permission of John Wiley & Sons, Inc.

9.5.3.2 Unsupervised classification

Unsupervised classification uses various **clustering** algorithms to determine the natural spectral groupings within an image (Lillesand & Kiefer 1994). One of these is the "K-means" approach, in which the analyst establishes the number of classes to be distinguished, and the algorithm then locates ("seeds") that number of classes within the multidimensional spectral space. Each pixel in the image is then assigned to the seed cluster that is closest to it within spectral space. After all pixels in the image have been assigned to a cluster, revised spectral characteristics are calculated for the comprehensive clusters, which are used to classify the image. This procedure is computationally intensive.

Another unsupervised classification approach is to use measures of image texture (see Section 3.2.4) to identify homogeneous areas. The mean of the first homogeneous window encountered in the image becomes the first cluster center, the mean of the second homogeneous window encountered becomes the second, and so forth.

After classes are identified by an unsupervised classification, they must be evaluated and named by the analyst. Merging of classes is generally required to reduce complexity and duplication. Both unsupervised and supervised classification tend to be iterative processes, whereby the results of the initial classification guide subsequent classifications until the final data layer is acceptable.

9.6 Integrating GIS and remote sensing

GIS and remote sensing capabilities are becoming increasingly integrated. We have seen that remote sensing is an important source of data for GIS analysis: imagery can be used with digital classification techniques to generate raster GIS layers, or with heads-up digitizing to generate vector GIS layers. Conversely, GIS data can serve as an important aid in image analysis. Urban areas, for example, are very difficult to accurately classify by digital image classification, because they are generally a mix of cover types: trees, lawns, buildings, pavement. GIS data are readily available for most urban areas, and can improve classification accuracy when used in combination with remotely sensed imagery (Harris & Ventura 1995). Similarly, elevation, slope, and aspect data can be used to aid forest classification in steep terrain, particularly if there are established relationships between topography and forest types (see Chapter 10). GIS data layers can be used to mask out portions of a remotely sensed image that might confuse digital classification of a feature of interest, such as using a GIS hydrography layer to mask lakes (low NIR reflectance) from an image to be used for classifying vegetation (high NIR reflectance). GIS data layers can be used as "ground truth" to check the accuracy of digital image classification.

Many image analysis systems can perform GIS procedures, particularly those associated with raster GIS. The ability to display GIS vectors superimposed over digital images is also becoming common, and will become more widely applied with the development of inexpensive georeferenced orthophotography. In time, the differences between image analysis and GIS will become less distinct, facilitating our ability to convey information from one to another.

CHAPTER 10

Modeling and GIS

Models provide ecologists with tools for extrapolating field measurements and integrating complex ecological information over space and time. This ability has become increasingly important as ecologists work at the broader scales of landscapes and the globe, because the scale of most ecological measurements is millimeters to meters. Furthermore, an individual measurement provides only one piece of ecological information and cannot account for the complexity, interaction, and dynamic nature of an entire ecosystem or landscape.

The integration of GIS with environmental models is emerging as a significant new area of GIS development and has been the topic of major conferences (Goodchild *et al.* 1993; Goodchild *et al.* 1996; NCGIA 1996). Reviews by Hunsaker *et al.* (1993) and Steyaert and Goodchild (1994) provide a comprehensive analysis of ecological models associated with GIS.

10.1 Types of models

The term "model" is used in a variety of ways. A model may be a **representation of data**, as a Digital Elevation Model (DEM – see Chapter 4) is a representation of topography. A **conceptual model** is an idea about how something functions, often depicted with block diagrams that show major systems, processes, and qualitative interrelationships among entities (Fig. 10.1). **Rule-based modeling** uses rules and numerical thresholds to interpret information represented in multiple data themes. **Mathematical modeling** involves the use of mathematical equations that may be implemented entirely within a GIS or executed in a separate model linked with a GIS.

Mathematical models are either **statistical** or **deterministic** (Steyaert 1993). Statistical models are based on empirical observations and contain one or more random variables, whereas a deterministic model does not have random variables. Both deterministic and statistical models can be either "steady-state" or "dynamic," the latter containing at least one term that is a function of time (Steyaert 1993). Environmental **simulation models** are mathematical models that mimic environmental processes (Law & Kelton 1982).

GIS has played various roles relative to modeling. **Cartographic modeling** (Tomlin 1990; Berry 1993) involves GIS analysis of spatial data with Boolean or mathematical operations, often to identify areas with

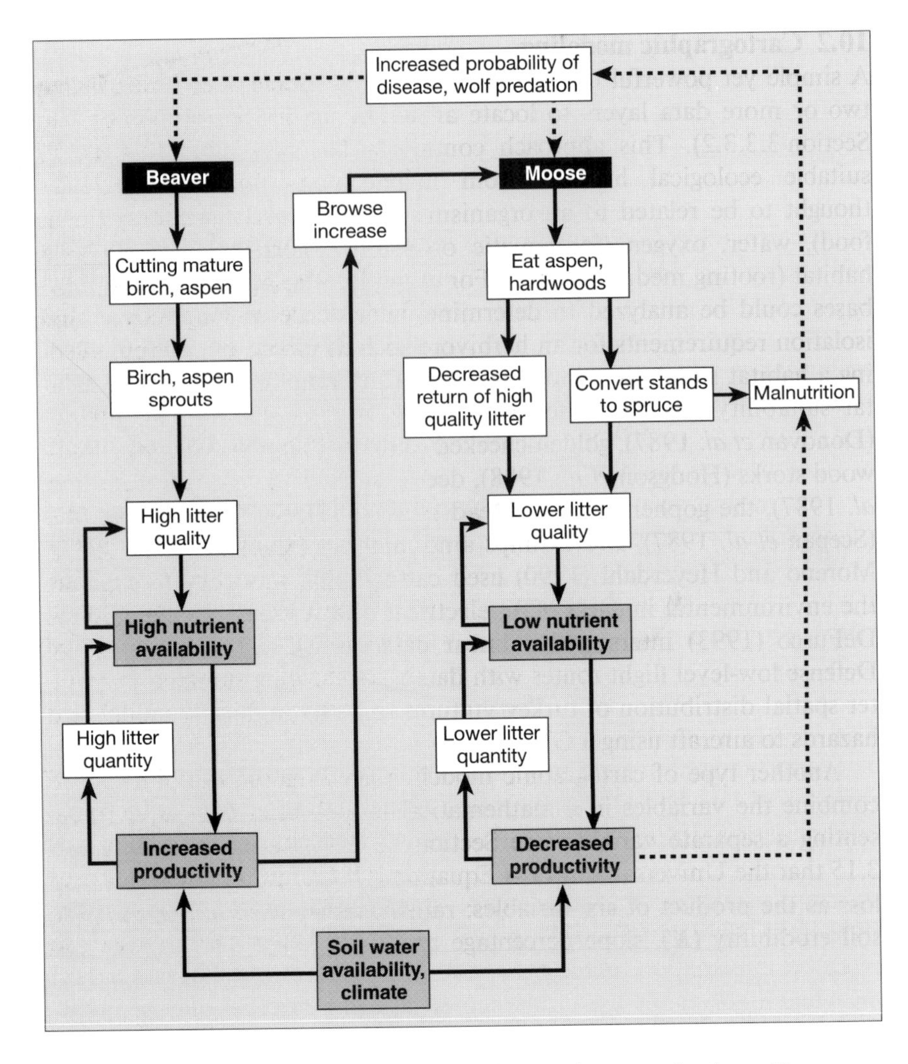

Fig. 10.1 Conceptual model of the interactions of moose, beaver, and wolves with vegetation and soil properties. From Johnston *et al.* 1993.

unique combinations of characteristics. In **statistical GIS modeling**, empirical relationships are developed between GIS-derived environmental characteristics (independent variables) and measures of ecological function (dependent variables). In **coupled GIS/simulation modeling**, GIS are used to derive input variables required by a simulation model; the GIS and model operations may be completely separate or may be tightly coupled by a software linkage that transparently passes data from one to the other. Examples of these different GIS/model uses are given below.

10.2 Cartographic modeling

A simple yet powerful use of GIS involves the Boolean combination of two or more data layers to locate areas having desired properties (see Section 3.3.3.2). This approach commonly has been used to identify suitable ecological habitats from mapped environmental variables thought to be related to an organism's survival, such as energy (light, food), water, oxygen (for aquatic organisms), nutrients, and physical habitat (rooting medium, space). For example, vegetation and road databases could be analyzed to determine coincidence of food, cover, and isolation requirements for an herbivore such as moose or caribou, yielding a habitat suitability map (Fig. 10.2). Cartographic modeling of habitat suitability has been used with many animal species: wild turkeys (Donovan *et al.* 1987), golden-cheeked warblers (Shaw & Atkinson 1988), wood storks (Hodgson *et al.* 1988), deer (Stenback *et al.* 1987; Tomlin *et al.* 1987), the gopher tortoise (Mead *et al.* 1988), the California condor (Scepan *et al.* 1987), and shrimps amd molluscs (Kapetsky *et al.* 1987). Moreno and Heyerdahl (1990) used cartographic modeling to evaluate the environmental impacts of an electrical transmission line on wildlife. DeFusco (1993) intersected a linear database of U.S. Department of Defense low-level flight routes with databases showing summer and winter spatial distribution of turkey vultures to cartographically model bird hazards to aircraft using a GIS.

Another type of cartographic modeling involves the use of a GIS to combine the variables in a mathematical model, each data layer representing a separate variable (see Section 3.3.3.3). Recall from Equation 3.15 that the Universal Soil Loss Equation (USLE) computes annual soil loss as the product of six variables: rainfall erosion index (R), inherent soil erodibility (K), slope percentage and slope length (LS), cover and

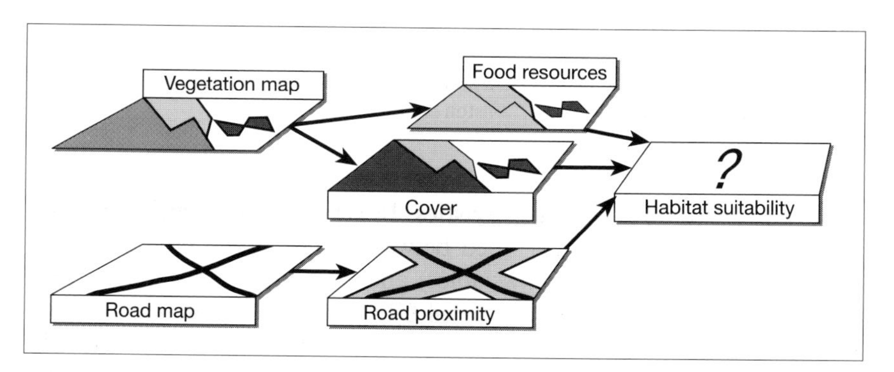

Fig. 10.2 Cartographic modeling of habitat suitability. The distributions of environmental variables known to affect organism abundance are used to predict the location of suitable habitat. From Goodchild *et al.* 1993. By permission of Oxford University Press.

management factor (C), and conservation practice factor (P). Several studies have used GIS to compute the USLE (and its successor, the Revised USLE) by deriving input variables from data layers for climate, soils, land management, and topography for each of the variables in the equation (Chrisman *et al.* 1986; Blaszczynski 1992). A similar cartographic modeling approach has been used to estimate agricultural nonpoint source (NPS) pollution potential for individual farms (Hession & Shanholtz 1988), the U.S. state of Pennsylvania (Hamlett *et al.* 1992), and the entire Mississippi River drainage basin (Gildea *et al.* 1986). Poiani and Bedford (1995) have evaluated the utility of such GIS-based modeling for estimating NPS loadings to wetlands.

A disadvantage of cartographic modeling is that the factors which determine the model outcome (i.e. soil loss rates, site suitability) must be known in advance. These relationships can be derived by statistical analysis, but more often they are developed subjectively, without rigorous testing. In such instances, cartographic modeling should be regarded as a spatial hypothesis, rather than scientific fact.

10.3 Rule-based modeling

Expert systems are computer systems that advise on or help solve real-world problems that would normally require a human expert's interpretation. Three types of rules are developed and programmed into the expert system: (i) database rules to evaluate numerical information; (ii) map rules to evaluate mapped categorical variables; and (iii) heuristic rules to evaluate the knowledge of domain experts (Coulson *et al.* 1991). Expert systems can be linked with a GIS to be made spatially explicit (Ripple & Ulshoefer 1987; Robinson & Frank 1987).

Several researchers have linked expert systems with GIS for habitat evaluation. Johnson *et al.* (1991) developed a linked GIS/expert system to evaluate bear habitat. Coulson and co-workers (1991) developed a linked GIS/expert system for predicting the hazard of southern pine beetle outbreaks, based on environmental factors (i.e. forest composition) and disturbance probabilities (i.e. stress potential, lightning strikes). Canadian workers (Lam & Swayne 1991) developed an expert system for modeling regional fish species richness.

Silveira *et al.* (1992) used rule-based modeling within the ERDAS GIS to investigate possible landscape/disturbance relationships in the Everglades ecosystem over time. Rules governing fire behavior, successional rates, and other relationships reported in published literature were generated using the ERDAS GISMO module. The resulting rule-based model was designed to simulate disturbance due to fire damage and post-disturbance vegetation succession.

In **inductive-spatial modeling**, a GIS "learns" relationships between datasets in the geographic database, developing rules based on analysis of the input data. Aspinall (1993) used this approach to model habitat suitability for red deer (*Cervus elaphus*) within the Grampian region of northeast Scotland. Databases of altitude, land cover, and accumulated frost were analyzed in combination with data on the distribution of red deer, recorded for each 1 × 1 km cell of the National Grid. The percentage of a habitat attribute in each grid cell occupied by red deer was divided by the percentage of that attribute in the area being considered, and transformed by adding one. The logarithm (base 10) of the resulting number provides an index of the importance of each category to the distribution of the species (Duncan 1983). Values less than 0.3 ($\log_{10}$ 2.0) reflect a category that is underrepresented, suggesting avoidance, and values greater than 0.3 suggest preference. Threshold values of these indices were used in a decision rule-base to classify red deer habitat preference (Aspinall 1993). Lees (1996) discusses various sampling strategies for machine learning that are applicable to linked GIS/expert systems.

Linked GIS/expert systems are increasingly being used to evaluate environmental risk (Fedra & Winkelbauer 1991; Lam *et al.* 1992; Fedra 1993). **Spatial Decision Support Systems** (SDSSs: Fig. 10.3) add the ability to recommend management solutions to environmental problems (Davis & Nanninga 1985; Fedra & Loucks 1985; White 1986; Johnston 1987; Parent & Church 1989; Fedra & Reitsma 1990; Maidment &

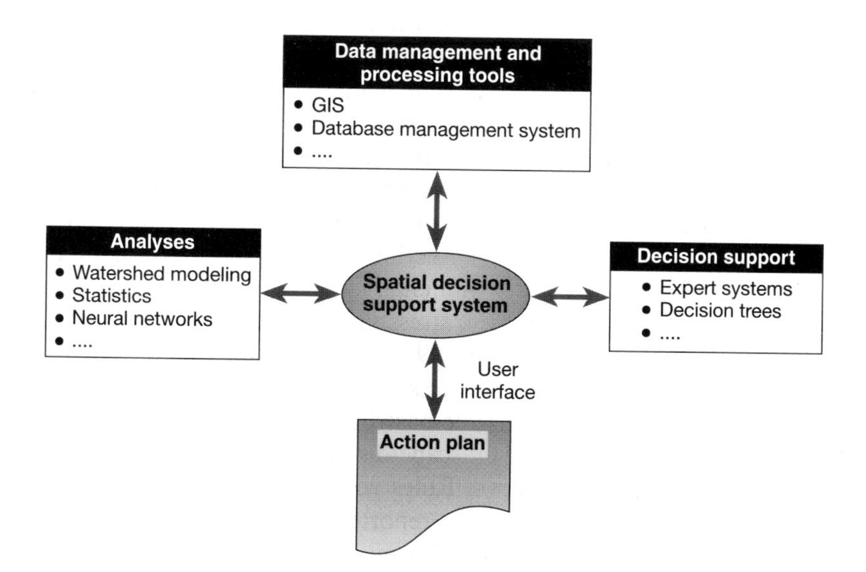

Fig. 10.3 Schematic of a spatial decision support shell. After Djokic 1996. By permission. *GIS and Environmental Modeling: Progress and Research Issues*, Goodchild *et al.* © 1996, GIS World Inc.

Djokic 1990; Kessell & Beck 1991; Djokic 1996). An SDSS can also help with visualizing the consequences of different management scenarios, which can aid in decision-making.

10.4 Statistical modeling

When the empirical relationships needed for model development are not known, a GIS can be used to assemble spatial data on landscape properties, derive new data that are syntheses of the originals, and statistically analyze those new data to determine the strength of interaction among them. The resulting empirical relationships can be used to predict gradients of habitat, net primary production, nutrient cycles, and other ecosystem properties across the landscape (Fig. 10.4). This approach would produce a whole landscape model as defined by Baker (1989).

In the studies described in this section, GIS-derived data were exported to a separate statistical package to generate the models. However, as the need by GIS users for more sophisticated statistical analyses has emerged, software has been developed specifically to link GIS with statistical packages, and more advanced statistical analysis routines have been built into GIS software.

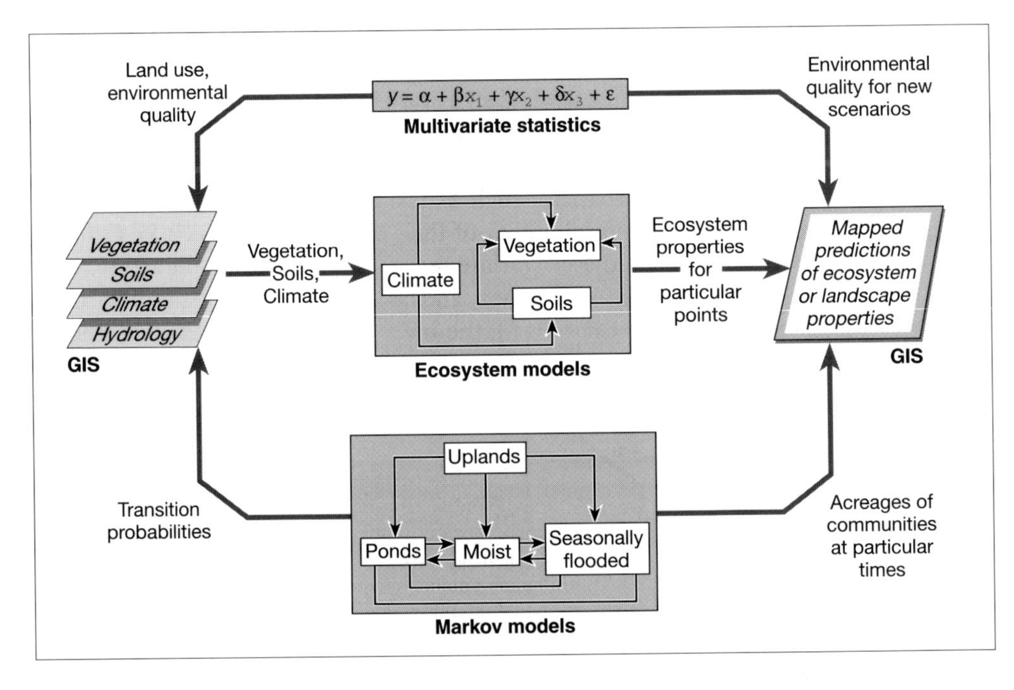

Fig. 10.4 Alternative methods of interfacing models with GIS. From Pastor & Johnston 1992.

10.4.1 Avoiding spatial bias in statistical modeling

A basic tenet of spatial statistics is that the attributes of locations that are close together are more likely to be similar than the attributes of locations that are far apart (see Chapter 7). If statistical modeling requires a spatial subset of data, it is important to minimize spatial bias (spatial autocorrelation) so as not to violate the requirement for independent observations imposed by classical statistics (Johnston 1992). Random sample selection is the method conventionally used to ensure independence of observations. Another sampling method is to choose sample points that are regularly spaced (i.e. lattice data: Cressie 1991) at a distance that meets an acceptable level of spatial autocorrelation. Methods for ensuring minimal spatial autocorrelation of a subset of GIS-derived data are discussed by Johnston (1992) and Anselin (1993).

10.4.2 Statistical models for continuous data

In a univariate statistical analysis, one independent variable (e.g. an environmental characteristic) is related to one dependent variable (e.g. an ecological function). Statistically validated relationships between organisms and environmental variables can be used to predict organism distributions from environmental conditions alone (static conditions), or organism response to environmental change (dynamic conditions).

In GIS-based univariate statistical modeling, dependent variables typically consist of field measurements performed at known locations (e.g. biomass, plant diversity, animal sightings), and independent variables are derived from a digital database containing continuous data (e.g. elevation, leaf area index). For example, some insect species are associated with mappable disturbances, such as fire or lightning strikes, which can be used to predict the probability of their occurrence (Fig. 10.5). Goward *et al.* (1986) developed a univariate regression model relating cumulative normalized difference vegetation index (NDVI) values derived from satellite imagery (see Chapter 6), the independent variable, to net annual

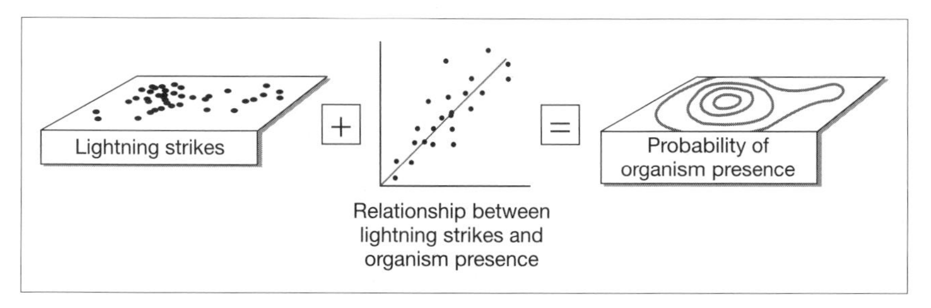

Fig. 10.5 Development of an empirical relationship between a population and a disturbance that affects it. The resulting equation is used to plot a probability surface of organism distribution. From Goodchild *et al.* 1993. By permission of Oxford University Press.

primary productivity for major North American biomes reported by Odum (1971), the dependent variable.

Statistical modeling has been used with aerial photograph-derived maps of vegetation disturbance to develop empirical relationships between animal populations and their disturbance effects. Broschart *et al.* (1989) used regression analysis to relate vegetation alteration by beaver dams to beaver colony density; GIS was used to summarize vegetation alteration from aerial photograph-derived maps for input into the statistical analysis. Ware and Maggio (1990) used analysis of covariance, a form of regression analysis that incorporates an additional element or covariate into the regression equations, to compare the linear expansion rate of oak wilt fungus in urban versus rural areas in the U.S. state of Texas; GIS analysis was used to determine the disease spread in eight different directions surrounding the calculated point of origin.

In multiple regression analysis, the dependent variable is statistically related to several independent variables, resulting in an equation of the form:

$$z = a_0 + a_1x_1 + a_2x_2 + a_3x_3 + \ldots \tag{10.1}$$

where a_0, a_1, a_2, etc. are the coefficients of the regression and x_1, x_2, x_3, etc. are the independent variables. Osborne and Wiley (1988) used multiple regression to empirically relate stream water quality (dependent variable) to three GIS-derived independent variables: watershed area, the ratio of urban to agricultural land area, and the ratio of forest to agricultural area. The latter two variables were analyzed within streamside buffer zones of four different widths (100, 200, 400, and 1000 ft), clipped as spatial subsets from a land use/cover map. The analysis indicated that the adverse water quality impact of increasing urbanization diminished with increasing distance from the stream channel.

Johnston (1992) used stepwise multiple regression to relate the percentage of trees per grid cell damaged by spruce budworm (dependent variable) to physical and vegetative site characteristics represented by a number of GIS data layers. The stepwise process selected distance from streams, percentage of pine cover, percentage of mature or overmature fir cover, and topographic slope as significant variables. Variables that were not selected by stepwise multiple regression included elevation, roads, a development code, windthrow damage, crown closure, and percentage of several vegetation types (immature fir, immature spruce, mature spruce, overmature spruce, hardwoods). Moore *et al.* (1993) used stepwise multiple regression to relate soil to topographic attributes derived from terrain analysis in a GIS, selecting only those attributes that significantly improved the regression at each step.

Multivariate statistical techniques coupled with GIS analyses provide the means for quickly compiling data, synthesizing these data, and developing predictive models to relate ecological functions to quantifiable landscape characteristics. Discriminant function analysis (DFA), principal components analysis (PCA), and other multivariate statistical techniques that can evaluate many variables simultaneously are useful in GIS modeling because they simplify the input data and identify key variables.

Johnston and co-workers (1988, 1990) used PCA to evaluate the effects of watershed characteristics in water quantity and quality, focusing on the effect of wetlands. Attributes of 15 watersheds in central Minnesota, U.S. were quantified in a GIS from digital maps of land use, soils, and topography. The 33 original landscape descriptors were reduced to eight principal components which explained 86% of the environmental variance. Using stepwise multiple regression, the scores of each of the watersheds for each of the principal components were then related to physical, chemical, and microbiological parameters of water draining from each watershed. The authors found that the position of wetlands upstream from the sampling station was related to their influence on water quality, a finding that would have been difficult to establish without the use of a GIS. A similar approach was used by the same researchers to evaluate the effects of wetlands on lake water quality (Detenbeck *et al.* 1993).

Discriminant function analysis (DFA) has been used with GIS-derived data to develop multivariate ecological models. Lowell (1991) used DFA with soils and topographic and historical vegetation data to develop a spatial model of ecological succession for an area composed of old fields, cedar glades, and oak–hickory forests. Clark *et al.* (1993) used DFA with telemetry data to model black bear habitat use.

10.4.3 Statistical models for categorical data

10.4.3.1 Expected versus observed outcomes

GIS databases frequently contain categorical data (see Section 2.2.3.3), which require different analysis techniques from those used with numerical data. One approach is to use statistical tests that compare expected versus observed outcomes. Pastor and Broschart (1990) used an electivity index to determine if forest communities were associated with particular soil types and slopes, and if they occurred in juxtaposition with each other in greater proportion than would be expected from their overall abundance in the landscape as a whole. The electivity index (Jacobs 1974) was computed for soil and slope classes as:

$$E_{ij} = \ln \frac{(r_{ij})(1 - p_i)}{(p_j)(1 - r_{ij})} \qquad (10.2)$$

where r is the proportion of community i on soil or slope class j, and p_j is the proportion of the landscape occupied by soil or slope class j. For the analysis of spatial adjacency of each cover type relative to all others, the form of the electivity index was:

$$E_{ij} = \ln \frac{(r_{ij})(1 - p_{ij})}{(p_{ij})(1 - r_{ij})} \tag{10.3}$$

where r_{ij} is the proportion of proximal cover type i in an arbitrarily selected 40-m buffer around cover type j, and p_{ij} is the proportion of proximal cover type in all 40-m wide buffers along all boundaries except those of cover type j. The electivity indices were tested against the chi-square distribution according to the formula:

$$\chi^2 = \frac{E_{ij}^2}{1/x_{ij} + 1/(m_j - x_{ij}) + 1/y_i + 1/(n_t - y_i)} \tag{10.4}$$

where x_{ij} is the area of cover type i in the buffers around cover type j, y_i is the area of cover type i in all 40-m wide buffers, m_j is the area of all cover types in the buffers around cover type j, and n_t is the area of all cover types in all buffers. The calculated χ^2 was compared with a χ^2 distribution with 1 degree of freedom at a significance level $p = 0.05$.

The use of chi-square analysis to compare expected versus observed occurrence has been used to develop statistical models of habitat preference for several wildlife species. Young *et al.* (1987) used chi-square and Bonferronni simultaneous confidence intervals with spotted owl radiotelemetry data to show that old growth was used significantly more than would be expected based on its proportion in the landscape, whereas none of the other seven habitat types analyzed were used more or less than proportional to availability. Agee *et al.* (1989) used chi-square analysis, *G*-statistics, and *t*-tests to determine if land cover classes, richness, and interspersion were significantly different in the vicinity of 91 grizzly bear (*Ursus arctos*) sightings than around 91 points randomly located on the same land cover database. Stoms *et al.* (1992) used a similar approach to evaluate habitat preference by the California condor (*Gymnogyps californianus*), and also evaluated the sensitivity of the resultant model to uncertainties in the input GIS data. Herr and Queen (1993) used chi-square analysis to determine the influence of zones of disturbance and non-disturbance (distance to roads, distance to buildings, distance to agriculture, width of undisturbed buffer) on sandhill crane (*Grus canadensis*) nesting.

Although each of the above wildlife habitat studies used similar statistical tests, they each used different GIS capabilities to define habitat zones. Young *et al.* (1987) defined "available habitat" as the proportion of cover types within home range polygons generated from the spotted owl radiotelemetry data, and compared it with "used habitat," the proportion of radio locations occurring in each habitat type (point data). Agee *et al.* (1989) evaluated habitat within a square matrix of pixels surrounding each grizzly bear sighting and randomly located point. Herr and Queen (1993) generated buffer zones around point (buildings), linear (roads), and polygon (agriculture) features to define potentially suitable crane nesting vegetation, from which they developed a model to estimate the number of nest sites. Thus, a variety of GIS techniques can be used to spatially subset a landscape for statistical analysis.

Logistic regression is a multiple regression technique suitable for use with categorical data. Unlike regression techniques used for continuous data, logistic regression compares the attributes of the locations where the phenomenon is present with those of the locations where the phenomenon is absent (Johnston 1992). A logistic regression takes the form:

$$z = \frac{1}{1 + \exp \Sigma n_i x_i} \qquad (10.5)$$

where n_i is the coefficient of the regression and x_i is an independent variable. Johnston (1992) used logistic regression to relate deer wintering yards (categorical, dependent variable) to various physical and vegetative characteristics represented by a number of GIS data layers, selecting variables for inclusion based on threshold values of two indices computed by the statistical analysis. The coefficients derived were used to create a deer probability surface. Pereira and Itami (1991) also used logistic multiple regression and Bayesian statistics (see below) to model the potential effects of a proposed observatory on the Mount Graham Red Squirrel and its habitat.

10.4.3.2 Bayesian statistics

Bayes' theorem provides a formal method for decision-making under conditions of uncertainty, and is a framework for combining relative values of being right or wrong (subjective probabilities) with the probabilities of being right or wrong (conditional probabilities) (Aspinall & Hill 1983). Milne *et al.* (1989) developed a Bayesian model of white-tailed deer (*Odocoileus virginianus*) habitat by comparing a map of deer wintering yards with 37 GIS databases of physical habitat (streams, roads, soils, glacial deposits) and forest cover variables. A one-way classification multivariate analysis of variance (MANOVA) was used to test for landscape variables that significantly contributed to differences in deer year presence or absence. Landscape variables were retained for use in Bayesian model development if they significantly reduced the unexplained variance in separate one-way analyses of variance; 12 soil type and forest composition variables were retained. Each pixel in the landscape was described by a vector, x, containing a list of the 12 variables present. Then, for each pixel the probability of each "state" of wildlife populations, w, was calculated, the two possible states being "deer habitat present" and "deer habitat absent."

According to Bayes' formula:

$$P(w_j|x) = \frac{p(x|w_j)\,P(w_i)}{p(x)} \qquad (10.6)$$

where

$$p(x) = \sum_{j=1}^{n} p(x|w_j)P(w_j) \qquad (10.7)$$

The term $p(x|w_j)$ is the state-conditional probability density function for x, $p(w_j)$ is the *a priori* probability of state w_j, and $P(w_j|x)$ is the *a posteriori* probability of each state, given the set of landscape variables present (Davis 1986). When equation (10.7) is used as a discriminant function, each pixel is assigned to whatever state has the highest *a posteriori* probability given the vector of landscape variables present. Milne *et al.* (1989) assigned pixels to the "deer habitat present" state if the *a posteriori* probability was greater than arbitrary threshold values, chosen such that the total amount of "deer habitat" constituted a 6.5, 15, or 30% proportion of the landscape. They compared the area of actual deer habitat with results of the Bayesian model, which predicted that deer would be present at sites where the probability of deer being present exceeds the probability that deer are absent. The number of predicted locations was smaller than the observed, indicating the conservative nature of the Bayesian estimate.

10.5 GIS coupled with mathematical models

The power of GIS becomes most apparent when they are coupled with simulation models of population, ecosystem, or global processes. This coupling is most successful with models that predict outcomes of processes such as succession, net primary production, and nutrient cycling, from parameters derived from maps or digital satellite data. Hypotheses are first formulated on how the behavior of organisms or ecosystems depends on their spatial relations with other organisms, systems, and the physical environment. The combinations of vegetation, soils, and other environmental variables are then identified. Spatial distribution, coincidence, or proximity of variables identified with the GIS can be input to computer models to examine the hypothesized consequences of spatial relations, In an iterative process, model predictions can be re-entered into the GIS to produce new maps of the predicted ecosystem properties along spatial gradients (Fig. 10.6). This iterative process can be used to simulate responses to new environmental conditions, such as climate change, by changing model input parameters (Cohen & Pastor 1991).

10.5.1 Population simulation models

Population growth at a particular location is affected by both intrinsic (e.g. birth rate, death rate, survivorship) and extrinsic factors (e.g. physical environment, interactions with other species) (Johnston 1993). Population ecologists have begun to build simulation models of population demographics that incorporate spatial elements, such as competitive

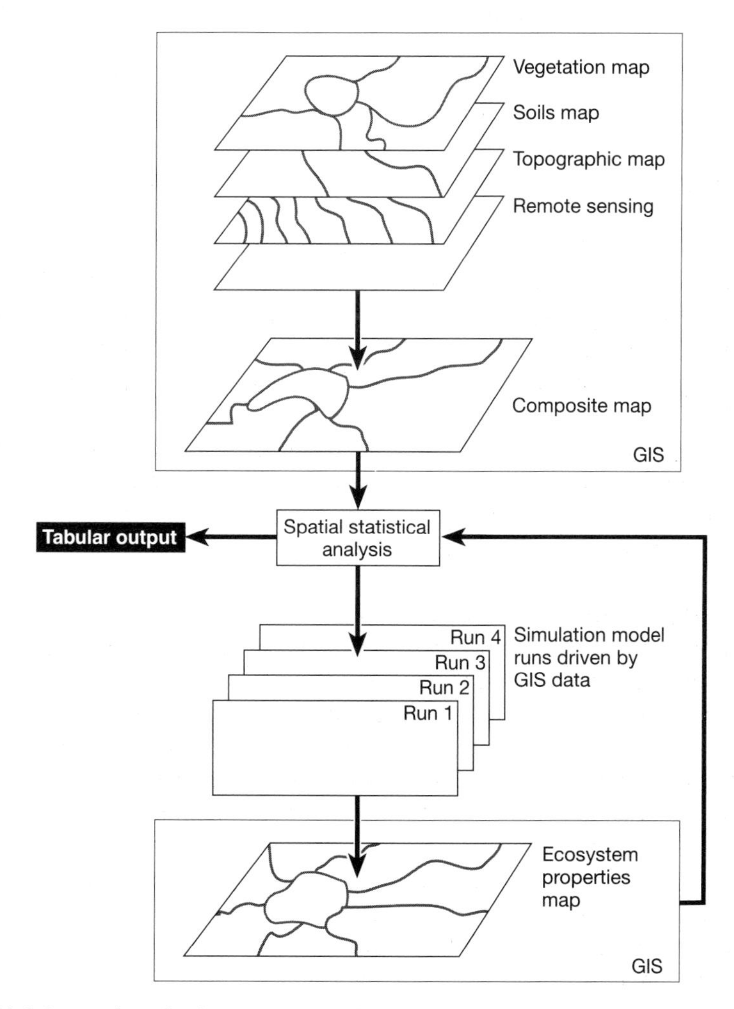

Fig. 10.6 Integration of a simulation model with a GIS.

interactions, dispersion, migration, and environmental patchiness (Sklar & Costanza 1991; Turner & Dale 1991). Incorporation of these spatial elements can improve the model results. For example, Parton and Risser (1980) found that a grassland simulation model would not work correctly without considering spatial distribution of the vegetation. Nachman (1987) designed a more stable predator–prey model by incorporating spatial realism and synergistic feedbacks such as environmental patchiness, dispersal characteristics, behavioral responses, and demographic stochasticity.

Polzer *et al.* (1991) developed a spatial landscape model using a database management system programming language to link a GIS with ZELIG, a forest growth model that simulates the establishment, growth, and mortality of trees within a grid of square plots, the size of which is determined by the model as a function of sun angle and the typical canopy height of the stand (Urban 1990). The developed software is capable of: (i) extracting input data and parameterizing the model; (ii) constructing a data link with the model; (iii) automating and executing the model; and (iv) analyzing and displaying results (Fig. 10.7). Feedback

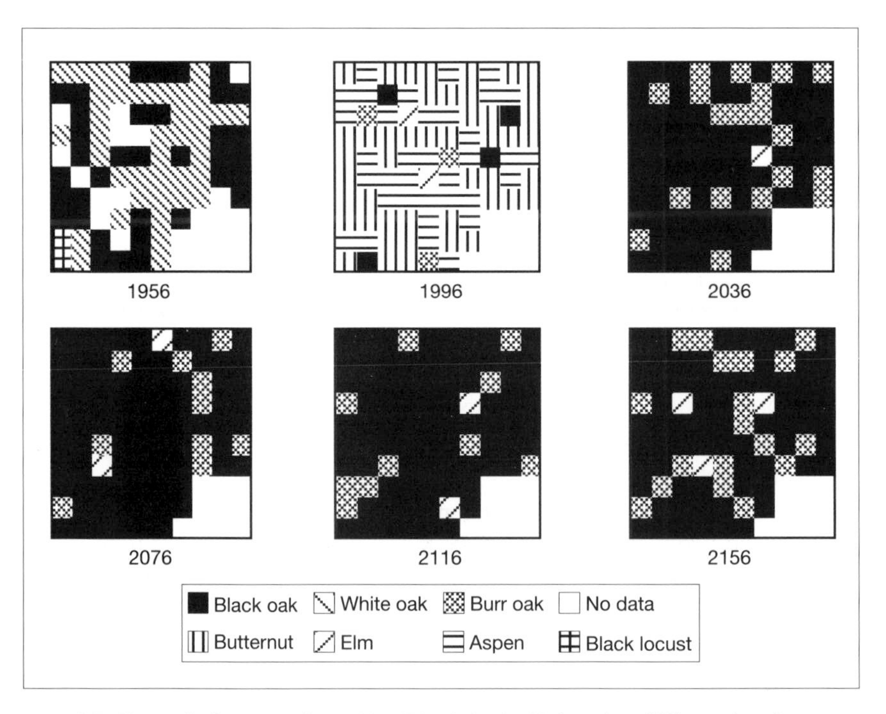

Fig. 10.7 Change in forest species at Noe Woods in the University of Wisconsin arboretum simulated by ZELIG. Reproduced with permission, the American Society for Photogrammetry and Remote Sensing. Polzer *et al.* (1991). In: *GIS/LIS '91 Proceedings*, 49–59.

from each model run is used as a source of input for successive model runs, to simulate long-term ecosystem changes. A variant of ZELIG uses a moving window to keep track of biomass and leaf area within adjacent cells, the results of which influence forest regeneration, growth, and mortality processes in the central cell of the moving window (Smith & Urban 1988). In this way, information from adjacent cells influences the outcome of simulated processes.

Incorporation of spatial elements into population models can also yield information about their influence on populations. Simulations with a population model incorporating organism dispersal demonstrated that a high dispersal and survival rate of dispersers was required to ensure survival in the presence of disturbance (Fahrig 1991). When the same model was applied to a simulated patchy landscape, the most important determinants of mean local population size were the fraction of organisms dispersing from the patches and the probability that the organisms would detect new patches. The linkage of such models with real landscapes in a GIS facilitates building and testing new population models that incorporate organism movements and environmental constraints.

10.5.2 Ecosystem and landscape simulation models

Whereas population models focus on organisms, ecosystem models and landscape models emphasize interactions among biotic and abiotic components. Ecosystem and landscape simulation models attempt to duplicate ecological function via coupled differential equations that describe key ecosystem and landscape processes. For example, the JABOWA and FORET forest models simulate the birth, growth, and death of individual trees based on deterministic, intrinsic stand variables (shading, crowding) and stochastic environmental variables (heat sums, temperature extremes, soil moisture) (Botkin *et al.* 1972; Shugart & West 1980).

GIS has been used with ecosystem and landscape models to derive input variables, spatially extrapolate results, and display results. Burke *et al.* (1990), for example, used a GIS to extract driving variables for the CENTURY grassland ecosystem model from climate and soils databases, run the model, and return the results (net primary production, soil organic carbon, net annual nitrogen mineralization, and net annual nitrous oxide production) to the GIS for display. This capability saved a substantial amount of model preparation time over conventional data input methods.

Forest growth models based on the ecology of individual trees are applicable at different spatial scales, because the cumulative growth of individual trees can be summed over a larger area to estimate biomass accrual of an entire forest. This principle was applied by Pastor and Post (1988), who used a version of FORET ("LINKAGES") to simulate the changing distribution of North American forest species under a projected climatic change caused by global warming. Their model assumed that species migrated to new sites to the extent that temperature and soil water availability were optimal and that growth was limited by temperature, water, nitrogen, or light, whichever was most restrictive under the altered climate. Forest biomass was predicted by species for 20 climatic stations across eastern North America, and extrapolated and displayed using a GIS (Pastor & Johnston 1992). Thus, the key to successfully extrapolating organism-scale models to landscapes or even continents is to focus on ecological properties that transcend spatial and temporal scales.

10.5.3 Spatially dynamic ecosystem models

Although the examples in the previous section have been made two-dimensional by using spatially explicit input parameters (e.g. Pastor & Post 1988; Polzer *et al.* 1991), they are still spatially static in the sense that the modeled units function independently of each other, with no fluxes of oganisms, materials, or information between them (Johnston *et al.* 1996). However, some ecological phenomena are stochastic and cannot be predicted by environmental variables alone. Examples are fire and weather events, which occur at unpredictable locations and times. These stochastic influences can be accounted for by incorporating Monte Carlo simulations in GIS models. A GIS can also reduce the spatial uncertainty associated with stochastic events by incorporating disturbance probabilities based on environmental characteristics, since the probability of a stochastic occurrence is not uniform at all locations. For example, some areas are more likely to be affected by fire or lightning strikes than others. A GIS can be used to evaluate landscape characteristics which ameliorate disturbance, and refine stochastic models accordingly. Predictive maps generated by the GIS can be compared with maps of actual occurrences to identify areas of disagreement due to stochastic influences.

A spatial version of the LINKAGES model, called S_L (Spatial Linkages), was recently developed to incorporate horizontal seed distribution into a forest growth model (Johnston *et al.* 1996). In S_L, a landscape of a pre-specified size is simulated. Plant species grow, reproduce, distribute seeds, and die on the landscape according to their tolerance to light, water, nitrogen conditions, seed production, and life-history characteristics. Simulations exhibit successional changes that are expected, produced mostly by competition for light and nitrogen, the availability of seeds, and the conditions for seed germination. The model was computationally intensive, and therefore could not be run in a GIS-software framework. Instead, a spatial landscape model was programmed in C++, and model output was directed to a video cassette recorder to produce a dynamic animation of changes in plant distribution and soil nitrogen across the landscape over time.

Hyman *et al.* (1991) incorporated animal movement into their herbivore model. The movements of model herbivores between cells are associated with a stimulus, a step size, a movement rule, and a maximum activity range. These same model constructs are used for all species and landscape scales; only the parameter values are changed. The model was used to simulate the effects of herbivores foraging independently of one another, as well as the repulsive effects of cell use by another individual. Spatially dynamic models have also been developed for moose foraging and limpet predation (Johnston *et al.* 1996).

None of the above spatially dynamic models were linked to a GIS, primarily because of their computational requirements. However, these examples point toward capabilities that could be incorporated in the future generation of GIS as computers become more powerful.

GIS data sources

Addresses, both surface mail and Internet, have a tendency to become obsolete. The following information is therefore provided as a guide, and is subject to change.

Digital Chart of the World

The Digital Chart of the World is available as a set of four CD ROMs for $200 by writing to:

In Canada:
Products & Services Division
Surveys, Mapping, and Remote
Sensing Sector
Energy, Mines & Resources Canada
615 Booth Street, Room 400
Ottawa, Ontario, K1A 0E4
Canada

In the U.S., Latin America, Asia, and Africa:
U.S. Geological Survey
ESIC-Open File Section
Box 25286
Federal Center
Mail Stop 517
Denver, CO 80225
U.S.A.

In Europe:
Chadwyck-Healey Ltd.
Cambridge Place
Cambridge CB2 INR
U.K.

In Australia:
The Manager
AUSMAP Data Unit
P.O. Box 2
Belconnen ACT 2617
Australia

Internet resources related to GIS

FAQs and resource catalogs

An **FAQ** provides on-line answers to "frequently asked questions" about a particular topic, and is an excellent starting point for finding information on the Internet. FAQs and other on-line documents particularly relevant to GIS are listed below.

"Frequently Asked Questions with Answers about Geographic Information Systems" © 1994, 1995, 1996 Lisa Nyman

Document may be freely distributed provided that copyright notice is provided, but permission is not granted for commercial use.

URLs: ftp://ftp.census.gov/pub/geo/gis-faq.txt
ftp://abraxas.adelphi.edu/pub/gis/FAQ
http://www.census.gov/geo/www/faq-index.html (hypertext version of the document)

To get this document via e-mail, send a mail message with a valid Internet "From" address (message content does not matter) to:
gis-faq-request@abraxas.adelphi.edu

To contribute to the FAQ, send an e-mail to: lnyman@census.gov

"Thoen's Web" © 1996 Bill Thoen and "Online Resources for Earth Scientists," © 1994 Bill Thoen

Thoen's Web and its ftp text predecessor are on-line catalogs containing a wealth of GIS-related information pertinent to ecological applications. The updated version of "Online Resources for Earth Scientists" is now embedded in Thoen's Web, and contains a large section on Biology.

URLs: http://www.csn.net/~bthoen
ftp://ftp.csn.org/COGS/ores.txt

"University of Edinburgh GIS Pointers Page"

Provides information about GIS, as well as a number of useful links to other servers worldwide. Provides access to Bruce Gittings' Digital Elevation Data catalog, which describes a large number of topographic and bathymetric datasets with local, regional, and global extent.

URL: http://www.geo.ed.ac.uk/home/gishome.html

Internet GIS data sources

"Australian Centre of the Asian Spatial Information and Analysis Network (ACASIAN)"

ACASIAN is an academic and applied research institute specializing in GIS databases for Asia and the former Soviet Union.

URL: http://www.asian.gu.edu.au/

"Australia Environment On-Line"
The Environmental Resources Information Network (ERIN) has been established in Australia to draw together and distribute information on the environment.
URL: http://kaos.erin.gov.au/erin.html

"ETOPO5 World Topographic Dataset"
These files contain worldwide bathymetric and elevation data in meters with a 5-minute by 5-minute latitude/longitude data density. The following ftp sites allow you to download the actual data:
URLs: ftp://walrus.wr.usgs.gov/pub/data/etopo5.northern.bat.Z
ftp://walrus.wr.usgs.gov/pub/data/etopo5.southern.bat.Z
ftp://ahab.rutgers.edu/pub/gridpak/etopo5

"Global Land Information System (GLIS)" U.S. Geological Survey
This is a telnet hypertext system of descriptive information about data sets for the Earth's land surfaces. Through GLIS, scientists can evaluate data sets such as Landsat or AVHRR satellite imagery, determine their availability, and place online requests for products. Although maintained by the U.S. government, its data are global in scope. Users without Internet access can install a graphical interface, PC-GLIS, on an IBM-compatible computer with modem to view images and other graphics.
URL: telnet://glis.cr.usgs.gov

"Global Resources Information Database (GRID) – Geneva"
GRID–Geneva is the UNEP/European office for environmental information and assessment activities.
URL: http://www.grid.unep.ch

"Manual of Federal Geographic Data Products," compiled by the Federal Geographic Data Committee (FGDC)
The manual describes U.S. Federal geographic data products that are national in scope and commonly distributed to the public, and is available via WWW on the Internet.
URL: http://info.er.usgs.gov/fgdc-catalog/title.html

"NAISMap Interactive GIS"
NAISMap is the first WWW-based GIS, which can be used to view and manipulate National Atlas spatial data layers and construct your own map of Canada.
URL: http://head-smashed-in.ccm.emr.ca/naismap/naismap.html

"National Cartographic Data Center (NCDC) On-Line Data Access"
The National Cartographic Data Center of the U.S. National Oceanic and Atmospheric Administration provides downloadable global climate and satellite data, including precipitation data from the U.S. NEXRAD radar system.
URL: http://www.ncdc.noaa.gov/homepg/online.html#DOWN

"Natural Resources Geographic Information System Laboratory"
The home page for the laboratory directed by the author. Contains descriptions, manuscripts, pictures, and downloadable demonstrations of ecological research applications utilizing GIS.
URL: http://www.nrri.umn.edu/gis.html

"United States Census Bureau"
Topographically Integrated Geographic Encoding and Referencing (TIGER) is the system and digital database developed to support the mapping needs of the Census Bureau's Decennial Census. The TIGER/Line®'95 database consists of a set of six CD ROMS showing streets, rivers, roads, and other features for the entire U.S. Ordering information is located at:
URL: http://www.census.gov

"United States Fish and Wildlife Service"
The FWS maintains a number of biological databases, including the National Wetlands Inventory:
URL: http://www.fws.gov/
Digital versions of NWI maps are available on-line at:
URL: http://www.nwi.fws.gov/

"United States Geological Survey Home Page" (USGS)
This is the main starting point on WWW for Earth and environmental science resources, for which this home page is the WWW registry. It contains on-line information about data from USGS, and also contains an extensive list of pointers to other servers that provide Earth and environmental science data managed by other agencies and universities.
URL: http://info.er.usgs.gov/
USGS geospatial data, information, public domain software, and related products are described at:
URL: http://www-nmd.usgs.gov/www/products/1product.html

Mailing lists and newsgroups

A **mailing list** automatically sends e-mail messages (including ones that you submit) to a group of subscribers with Internet addresses. Once you subscribe to a mailing list, you receive a copy of every e-mail message sent to the list. A **newsgroup** is a special interest news distribution scheme that allows individuals with access to it to read and post articles. Advantages of newsgroups over mailing lists include the sorting of articles on related topics, the control of distribution of posted articles to hierarchical levels, the ability to cancel an article even after it has been distributed, and automatic expiration of dated articles.

The GIS-L mailing list consists of a large number of people interested in GIS from around the world. GIS-L generates 50 messages or more a day, so access may be more convenient via its Usenet newsgroup:

comp.infosystems.gis.

Both GIS-L and its newsgroup contain the same information, and both allow individuals to read and post articles. To subscribe to GIS-L, send an e-mail message saying:

SUBSCRIBE GIS-L <your_name>

to the server: LISTSERVE@URISA.ORG

Within a day or so, you will receive an e-mail message from the server confirming your subscription and providing you with additional information and instructions about using the GIS-L list. Items to be posted to the list should be sent to GIS-L@URISA.ORG or posted to comp.infosystems.gis

References

Acevedo, M.F., Urban, D.L. & Ablan, M. (1996) Landscape scale forest dynamics: GIS, gap and transition models. In: Goodchild, M.R., Steyaert, L.T., Parks, B.O. *et al.* (eds) *GIS and Environmental Modeling: Progress and Research Issues*, 181–185. GIS World Books, Boulder.

Agee, J.K., Stitt, S.C.F., Nyquist, M. & Root, R. (1989) Geographic analysis of historical grizzly bear sightings in the North Cascades. *Photogrammetric Engineering and Remote Sensing*, **55**, 1637–1642.

Allen, T.F.H. & Starr, T.B. (1982) *Hierarchy: Perspectives for Ecological Complexity*. Chicago Press, Chicago, IL.

Al-Taha, K. & Barrera, R. (1990) Temporal data and GIS: an overview. In: *GIS/LIS'90 Proceedings, Anaheim, CA, Volume 1*, 244–254, American Society for Photogrammetry and Remote Sensing, Bethesda, MD.

Anderson, J.R., Hardy, E.E., Roach, J.T. & Witmer, R.E. (1976) *A Land Use and Land Cover Classification System for Use with Remote Sensor Data*. U.S. Geological Survey Professional Paper 964, U.S. Government Printing Office, Washington, D.C.

Anderson, T.W. & Goodman, L.A. (1957) Statistical inference about Markov chains. *The Annals of Mathematical Statistics*, **28**, 89–110.

Anselin, L. (1993) Discrete space autoregressive models. In: Goodchild, M.R., Parks, B.O. & Steyaert, L.T. (eds) *Environmental Modeling with GIS*, 454–469. Oxford University Press, New York, NY.

Antenucci, J.C., Brown, K., Croswell, P.L. & Kevany, M.J. (1991) *Geographic Information Systems: A Guide to the Technology*. Van Nostrand Reinhold, New York, NY.

Aronoff, S. (1989) *Geographic Information Systems: A Management Perspective*. WDL Publications, Ottawa.

Arthur, J.W., Roush, T., Thompson, J.A., Puglisi, F.A., Richards, C., Host, G.E. & Johnson, L.B. (1996) Evaluation of watershed quality in the Saginaw River Basin. *U.S. Environmental Protection Agency Publication EPA/600/R-95/153*. U.S. EPA Office of Research and Development, Washington, D.C.

Aspinall, P.J. & Hill, A.R. (1983) Clinical inferences and decisions — I. Diagnosis and Bayes' theorem. *Ophthalmic Physiology Optician*, **3**, 295–304.

Aspinall, R.J. (1993) Use of geographic information systems for interpreting land-use policy and modeling effects of land-use change. In: Cousins, S.H., Haines-Young, R. & Green, D. (eds) *Landscape Ecology and Geographic Information Systems*, 223–236. Taylor & Francis, London.

Aspinall, R.J. (1994) GIS and spatial analysis for ecological modelling. In: Michener, W.K., Brunt, J.W. & Stafford, S.G. (eds) *Environmental Information Management and Analysis: Ecosystem to Global Scales*, 377–396. Taylor & Francis, London.

Aspinall, R.J. & Pearson, D.M. (1996) Data quality and spatial analysis: analytical use of GIS for ecological modeling. In: Goodchild, M.F., Steyaert, L.T., Parks, B.O. *et al.* (eds) *GIS and Environmental Modeling: Progress and Research Issues*, 35–38. GIS World Books, Fort Collins, CO.

Aspinall, R.J. & Veitch, N. (1993) Habitat mapping from satellite imagery and wildlife survey data using a Bayesian modeling procedure in a GIS. *Photogrammetric Engineering & Remote Sensing*, **59**, 537–543.

Atkinson, P.M. & Harrison, A.R. (1993) Optimal sampling and estimation of land use. In:

GIS/LIS'93 Proceedings, Minneapolis, MN, Volume 1, 30–39. American Society for Photogrammetry and Remote Sensing, Bethesda, MD

August, P., Michaud, J., Labash, C. & Smith, C. (1994) GPS for environmental applications: accuracy and precision of locational data. *Photogrammetric Engineering and Remote Sensing*, **60**, 41–45.

Baker, W.L. (1989) A review of models of landscape change. *Landscape Ecology*, **2**, 111–133.

Baker, W.L. & Cai, Y. (1992) The r.le programs for multiscale analysis of landscape structure using the GRASS geographical information system. *Landscape Ecology*, **7**, 291–302.

Baker, W.L., Cai, Y., Musleh, S. & Bucher, J. (1993) *The r.le Programs. Version 2.0.5 Beta*. Department of Geography and Recreation, University of Wyoming, Laramie, WY.

Baker, W.L., Egbert, S.L. & Frazier, G.F. (1991) A spatial model for studying the effects of climatic change on the structure of landscapes subject to large disturbances. *Ecological Modeling*, **56**, 109–125.

Band, L.E. (1986) Topographic partitioning of watersheds with digital elevation models. *Water Resources Research*, **22**, 15–24.

Band, L.E., Peterson, D.L., Running, S.W. *et al.* (1991) Forest ecosystem processes at the watershed scale: basis for distributed simulation. *Ecological Modeling*, **56**, 171–196.

Barry, C.M. (1993) Hartwell Lake, PCBs, and GPS. *GPS World*, **4**(5), 44–46.

Bartz, K.L., Kershner, J.L., Ramsey, R.D. & Neale, C.M.U. (1992) Assessing riparian vegetation using multispectral, airborne videography: a new resource management tool. In: *Proceedings of the Fourth Forest Service Remote Sensing Applications Conference*, 319–327. American Society for Photogrammetry and Remote Sensing, Bethesda, MD.

Baskin, R.L. (1992) GPS finds its place on the lake. *GPS World*, **3**(5), 27.

Beals, E.W. (1969) Vegetational change along altitudinal gradients. *Science*, **165**, 981–985.

Befort, W. (1986) Large-scale sampling photography for forest habitat-type identification. *Photogrammetric Engineering and Remote Sensing*, **52**, 101–108.

Bell, K.R., Blanchard, B.J., Schmugge, T.J. & Witczak, M.W. (1980) Analysis of surface moisture variations within large-field sites. *Water Resources Research*, **16**, 796–810.

Berger, P., Meysembourg, P., Salés, J. & Johnston, C.A. (1996) Towards a virtual reality interface for landscape visualization. *Proceedings of the Third International Conference/Workshop on Integrating GIS and Environmental Modeling*. National Center for Geographic Information and Analysis, Santa Barbara, CA (CD ROM).

Berry, J.K. (1993a) *Beyond Mapping: Concepts, Algorithms, and Issues in GIS*. GIS World Books, Fort Collins, CO.

Berry, J.K. (1993b) Cartographic modeling: the analytical capabilities of GIS. In: Goodchild, M.R., Parks, B.O. & Steyaert, L.T. (eds) *Environmental Modeling with GIS*, 58–74. Oxford University Press, New York, NY.

Bertram, T.E. & Cook, A.E. (1993) Satellite imagery and GPS-aided ecology. *GPS World*, **4**(10), 48–53.

Bian, L., Sun, H., Bledgett, C. *et al.* (1996) An integrated interface system to couple the SWAT model and ARC/INFO. *Proceedings of the Third International Conference/Workshop on Integrating GIS and Environmental Modeling*. National Center for Geographic Information Analysis, Santa Barbara, CA (CD ROM).

Bie, S.W & Beckett, P.H.T. (1973) Comparison of four independent soil surveys by air-photo interpretation, Paphos area (Cyprus). *Photogrammetria*, **29**, 189–202.

Blaszczynski, J. (1992) Regional soil loss prediction utilizing the RUSLE/GIS interface. In: Johnson, A.I., Pettersson, C.B. & Fulton, J.L. (eds) *Geographic Information Systems (GIS) and Mapping – Practices and Standards*, 122–131. American Society for Testing and Materials, Philadelphia, PA.

Bobbe, T.J. (1992) Real-time differential GPS for aerial surveying and remote sensing. *GPS World*, **3**(7), 18–22.

Bobbe, T.J., Ishikawa, P., Jr., Reutebuch, S. & Hoppus, M. (1992) Creating a riparian – vegetation GIS data base from NASA ER-2 high altitude color infrared stereomodels. In: *ASPRS/ACSN/RT 92 Technical Papers, Volume 3*, 45–54. American Society for Photogrammetry and Remote Sensing, Bethesda, MD.

Bobbe, T.J., Alban, J.A., Ishikawa, P., Jr. & Myhre, R.J. (1994) An evaluation of narrow-band multispectral video imagery for monitoring forest health. In: *Proceedings of the Fifth Forest Service Remote Sensing Applications Conference*, 191–197. USDA, Portland, OR.

Bolstad, P.V. (1991) Positional uncertainty in natural resource spatial data: the contribution of camera tilt and terrain relief. In: *GIS/LIS'91 Proceedings, Volume 2*, 874–881. American Society for Photogrammetry and Remote Sensing, Bethesda, MD.

Botkin, D.B., Janak, J.F. & Wallis, J.R. (1972) Some ecological consequences of a computer model of forest growth. *Journal of Ecology*, **60**, 849–872.

Brandt, S.B., Mason, D.M., Goyke, A., Hartman, K.J., Kirsch, J.M. & Luo, J. (1996) Spatial modeling of fish growth: underwater views of the aquatic habitat. In: Goodchild, M.R., Steyaert, L.T., Parks, B.O. *et al.* (eds) *GIS and Environmental Modeling: Progress and Research Issues*, 225–229. GIS World Books, Boulder, CO.

Brimicombe, A.J. (1993) Combining positional and attribute uncertainty using fuzzy expectation in a GIS. In: *GIS/LIS'93 Proceedings, Minneapolis, MN, Volume 2*, 72–81. American Society for Photogrammetry and Remote Sensing, Bethesda, MD.

Brimicombe, A.J., Barlett, J.M. (1996) Linking GIS with hydraulic modeling for flood risk assessment: the Hong Kong approach. In: Goodchild, M.R., Steyaert, L.T., Parks, B.O. *et al.* (eds) *GIS and Environmental Modeling: Progress and Research Issues*, 165–168. GIS World Books, Fort Collins, CO.

Broschart, M.R., Johnston, C.A. & Naiman, R.J. (1989) Predicting beaver colony density in boreal landscapes. *Journal of Wildlife Management*, **53**, 929–934.

Brown, J.F., Loveland, T.R., Merchant, J.W., Reed, B.C. & Ohlen, D.O. (1993) Using multisource data in global land-cover characterization: concepts, requirements, and methods. *Photogrammetric Engineering and Remote Sensing*, **59**, 977–987.

Brunt, J.W. & Conley, W. (1990) Behavior of a multivariate algorithm for ecological edge detection. *Ecological Modeling*, **49**, 179–203.

Buol, S.W., Hole, F.D. & McCracken, R.J. (1980) *Soil Genesis and Classification*, 2nd edn. Iowa State University Press, Ames, IA.

Burgess, T.M. & Webster, R. (1980a) Optimal interpolation and isarithmic mapping of soil properties. I. The semi-variogram and punctual kriging. *Journal of Soil Science*, **31**, 315–331.

Burgess, T.M. & Webster, R. (1980b) Optimal interpolation and isarithmic mapping of soil properties. II. Block kriging. *Journal of Soil Science*, **31**, 333–341.

Burgess, T.M., Webster, R. & McBratney, A.B. (1981) Optimal interpolation and isarithmic mapping of soil properties. IV. Sampling strategy. *Journal of Soil Science*, **31**, 643–659.

Burke, I.C., Schimel, D.S., Yonker, C.M., Parton, W.J., Joyce, L.A. & Lauenroth, W.K. (1990) Regional modeling of grassland biogeochemistry using GIS. *Landscape Ecology*, **45**, 45–54.

Burrough, P.A. (1984) The application of fractal ideas to geophysical phenomena. *Bulletin of the Institute of Mathematics and its Applications*, **20**(3/4), 36–42.

Burrough, P.A. (1986) *Principles of Geographical Information Systems for Land Resources Assessment*. Monographs on Soil and Resources Survey No. 12. Oxford University Press, Oxford.

Burrough, P.A. (1989) Fuzzy mathematical methods for soil survey and land evaluation. *Journal of Soil Science*, **40**, 477–492.

Burrough, P.A., van Rijn, R. & Rikken, M. (1996) Spatial data quality and error analysis issues: GIS functions and environmental modeling. In: Goodchild, M.F., Steyaert, L.T.,

Parks, B.O. *et al.* (eds) *GIS and Environmental Modeling: Progress and Research Issues*, 29–34. GIS World Books, Fort Collins, CO.

Carter, J.R. (1984) *Computer Mapping, Progress in the '80s*. Association of American Geographers, Washington, D.C.

Carter, J.R. (1988) Digital representations of topographic surfaces. *Photogrammetric Engineering and Remote Sensing*, **54**, 1577–1580.

di Castri, F., Hansen A. & Holland, M.M. (1988) A new look at ecotones: emerging international projects on landscape boundaries. *Biology International*, **17**, Special Issue.

Caswell, H. (1989) *Matrix Population Models*. Sinauer Associates, Sunderland, MA.

Cheves, M. (1997) NGS releases new geoid model. *Professional Surveyor*, **17**(1), 49–51.

Chrisman, N.R. (1994) Metadata required to determine the fitness of spatial data for use in environmental analysis. In: Michener, W.K., Brunt, J.W. & Stafford, S.G. (eds) *Environmental Information Management and Analysis: Ecosystem to Global Scales*, 177–190. Taylor & Francis, London.

Chrisman, N.R., Mezera, D.F., Moyer, D.D., Niemann, B.J., Jr., Sullivan, J.G. & Vonderohe, A.P. (1986) Soil erosion planning in Wisconsin: an application and evaluation of a multipurpose land information system. In: *Technical Papers, 1986 ACSM-ASPRS Annual Convention*, 240–249. American Society for Photogrammetry and Remote Sensing, Falls Church, VA.

Chuvienco, E. & Martín, M.P. (1994) Global fire mapping and fire danger estimation using AVHRR images. *Photogrammetric Engineering and Remote Sensing*, **60**(5), 563–570.

Clark, C.A., Cate, R.B., Trenchard, M.H., Boatright, J.A. & Bizzell, R.M. (1986) Mapping and classifying large ecological units. *BioScience*, **36**, 476–478.

Clark, I. (1980) The semivariogram. In: *Geostatistics*, 17–28. McGraw-Hill, New York, NY.

Clark, J.D., Dunn, J.E. & Smith, K.G. (1993) A multivariate model of female black bear habitat use for a geographic information system. *Journal of Wildlife Management*, **57**, 519–526.

Clarke, K.C. & Olsen, F. (1996) Refining a cellular automaton model of wildfire propagation and extinction. In: Goodchild, M.R., Steyaert, L.T., Parks, B.O. *et al.* (eds) *GIS and Environmental Modeling: Progress and Research Issues*, 333–338. GIS World Books, Fort Collins, CO.

Cliff, A.D. & Ord, J.K. (1973) *Spatial Processes, Models, and Applications*. Pion, London.

Cliff, A.D., Haggett, P., Ord, J.K., Bassett, K.A. & Davies, R.B. (1975) *Elements of Spatial Structure: A Quantitative Approach*. Cambridge University Press, Cambridge.

Cline, M.G. (1949) Basic principles of soil classification. *Soil Science*, **67**, 81–91.

Cline, M.G. (1963) Logic of the new system of soil classification. *Soil Science*, **96**, 17–22.

Cohen, Y. & Pastor, J. (1991) The responses of a forest model to serial correlations of global warming. *Ecology*, **72**, 1161–1165.

Colwell, R.N. (ed.) (1983) *The Manual of Remote Sensing*. American Society of Photogrammetry, Falls Church, VA.

Congalton, R.G., Green, K. & Teply, J. (1993) Mapping old growth forests on national forest and park lands in the Pacific northwest from remotely sensed data. *Photogrammetric Engineering and Remote Sensing*, **59**(4), 529–535.

Cottam, G. & Curtis, J.T. (1956) The use of distance measures in phytosociological sampling. *Ecology*, **37**, 451–460.

Coulson, R.N., Lovelady, C.N., Flamm, R.O., Spradling, S.L. & Saunders, M.C. (1991) Intelligent geographic information systems for natural resource management. In: Turner, M.G. & Gardner, R.H. (eds) *Quantitative Methods in Landscape Ecology*, 153–172. Springer-Verlag, New York, NY.

Cousins, S.H., Haines-Young, R. & Green, D. (eds) (1993) *Landscape Ecology and Geographic Information Systems*. Taylor & Francis, London.

Cowardin, L.M., Arnold, P.M., Shaffer, T.C., Pywell, H.R. & Miller, L.D. (1989) Duck numbers estimated from ground counts, MOSS map data, and aerial video. In: Scurry,

J.D. (ed.) *Proceedings of the 5th National MOSS Users Workshop*, 205–219. U.S. Fish and Wildlife Service, Slidell, LA.

Cowardin, L.M., Carter, V., Golet, F.C. & LaRoe, E.T. (1979) Classification of wetlands and deepwater habitats of the United States. In: *U.S. Fish and Wildlife Service Pub. FWS/OBS-79/31*, U.S. Fish and Wildlife Service, Washington, D.C.

Cowen, D.J. (1983) Rethinking DIDS: the next generation of interactive color mapping systems. *Cartographica*, **21**, 89–92.

Cowen, D.J. (1988) GIS vs. CAD versus DBMS: what are the differences? *Photogrammetric Engineering and Remote Sensing*, **54**, 1551–1555.

Craig, W.J. (1985) The Minnesota Land Management Information System ten years later. In: Hocking, P.J. (ed.) *Proceedings of the 1985 Conference of the Australian Urban and Regional Information Systems Association*, 279–290. Australian Urban and Regional Information Systems Association, Canberra.

Cressie, N.A.C. (1991) *Statistics for Spatial Data*. Wiley, New York, NY.

Curtis, J.T. (1959) *The Vegetation of Wisconsin*. University of Wisconsin Press, Madison, WI.

Daly, C. & Taylor, G.H. (1996) Development of a new Oregon precipitation map using the PRISM model. In: Goodchild, M.R., Steyaert, L.T., Parks, B.O. *et al.* (eds) *GIS and Environmental Modeling: Progress and Research Issues*, 91–92. GIS World Books, Fort Collins, CO.

Dana, P.H. (1995) *The Geographer's Craft Project*. Department of Geography, University of Texas, Austin, TX.

Davis, B.A., George, J.R. & Marx, R.W. (1992) TIGER/SDTS: standardizing an innovation. *Cartography and Geographic Information Systems*, **19**, 321–327.

Davis, F.W. & Goetz, S. (1990) Modeling vegetation pattern using digital terrain data. *Landscape Ecology*, **4**, 69–80.

Davis, F.W. & Simonett, D. (1991) Remote sensing and GIS. In: Maguire, D.J., Goodchild, M.F. & Rhind, D.W. (eds) *Geographic Information Systems: Principles and Applications*. Longman Scientific and Technical, London.

Davis, F.W., Quattrochi, D.A., Ridd, M.K. *et al.* (1991) Environmental analysis using integrated GIS and remotely sensed data: some research needs and priorities. *Photogrammetric Engineering and Remote Sensing*, **57**, 689–697.

Davis, J.C. (1986) *Statistics and Data Analysis in Geology*. 2nd edn. Wiley, New York, NY.

Davis, J.R. & Nanninga, P.M. (1985) GEOMYCIN: towards a geographic expert system for resource management. *Journal of Environmental Management*, **21**, 377–390.

Defense Mapping Agency (DMA) (1992) *VPF VIEW 1.0 User's Manual for the Digital Chart of the World for Use with the Disk Operating System (DOS), Defense Mapping Agency, Washington, D.C.*

DeFusco, R.P. (1993) Modeling bird hazards to aircraft: a GIS application study. *Photogrammetric Engineering and Remote Sensing*, **59**, 1481–1496.

Delcourt, P.A. & Delcourt, H.R. (1992) Ecotone dynamics in space and time. In: di Castri, F. & Hansen, A. (eds) *Landscape Boundaries: Consequences for Biotic Diversity and Ecological Flows*, 19–54 Springer-Verlag, New York, NY.

Delhomme, J.P. (1978) Kriging in the hydrosciences. *Advances in Water Research*, **1**, 251–266.

Detenbeck, N.E., Johnston, C.A. & Niemi, G.J. (1993) Wetland effects on lake water quality in the Minneapolis/St. Paul metropolitan area. *Landscape Ecology*, **8**, 39–61.

Deutsch, C.V. & Journel, A.G. (1992) *GSLIB: Geostatistical Software Library and User's Guide*. Oxford University Press, New York, NY.

Djokic, D. (1996) Toward a general purpose spatial decision support system using existing technologies. In: Goodchild, M.R., Steyaert, L.T., Parks, B.O. *et al.*, (eds) *GIS and Environmental Modeling: Progress and Research Issues*, 353–356. GIS World Books, Fort Collins, CO.

Donovan, M.L., Rabe, D.L. & Olson, C.E., Jr. (1987) Use of geographic information systems to develop habitat suitability models. *Wildlife Society Bulletin*, **15**, 574–579.

Douglas, D.H. & Peuker, T.K. (1973) Algorithms for the reduction of the number of points required to represent a digitized line, or its caricature. *Canadian Cartography*, **10**, 112–122.

Downes, R.G. & Beckwith, R.S. (1951) Studies in the variation of soil reaction. Field variations at Barooga, N.S.W. *Australian Journal of Agricultural Research*, **2**, 60–72.

Dozier, J. (1980) A clear-sky spectral solar radiation model for snow-covered mountainous terrain. *Water Resources Research*, **16**, 709–718.

Drake, P. & Luepke, D. (1991) GPS for forest fire management and cleanup. *GPS World*, **2**(8), 42–46.

Druss, M (1992) Recovering history with GPS. *GPS World*, **3**(4), 32–37.

Dubayah, R., & Rich, P.M. (1996) GIS-based solar radiation modeling. In: Goodchild, M.R., Steyaert, L.T., Parks, B.O. *et al.* (eds) *GIS and Environmental Modeling: Progress and Research Issues*, 129–134. GIS World Books, Fort Collins, CO.

Dubrule, O. (1983) Two methods with different objectives: splines and kriging. *Mathematical Geology*, **15**, 245–255.

Duguay, C.R. & Walker, D.A. (1996) Environmental modeling and monitoring with GIS: Niwot Ridge Long-Term Ecological Research Site. In: Goodchild, M.R., Steyaert, L.T., Parks, B.O. *et al.* (eds) *GIS and Environmental Modeling: Progress and Research Issues*, 219–223. GIS World Books, Fort Collins, CO.

Duncan, P. (1983) Determinants of the use of habitat by horses in a Mediterranean wetland. *Journal of Animal Ecology*, **52**, 93–109.

Dunn, C.P., Sharpe, D.M. Guntenspergen, G.R., Stearns, F. & Yang Z. (1991) Methods for analyzing temporal changes in landscape pattern. In: Turner, M.G. & Gardner R.H. (eds) *Quantitative Methods in Landscape Ecology*, 173–198. Springer-Verlag, New York, NY.

Eastman, J.R. (1992a) *IDRISI: A Grid Based Geographic Analysis System*. Version 4.0, Clark University, Graduate School of Geography, Worcester, MA.

Eastman, J.R. (1992b) Time series map analysis using standardized principal components. In: *Proceedings, ASPRS/ACSM/RT '92 Convention on Monitoring and Mapping Global Change, Volume 1*, 195–204. American Society for Photogrammetry and Remote Sensing, Bethesda, MD.

Eastman, J.R. & Fulk, M. (1993) Long sequence time series evaluation using standardized principal components. *Photogrammetric Engineering and Remote Sensing*, **59**, 991–996.

Ehlers, M., Edwards, G. & Bedard, Y. (1989) Integration of remote sensing with geographic information systems: a necessary evolution. *Photogrammetric Engineering and Remote Sensing*, **55**, 1619–1627.

Eidenshink, J.C. (1992) The 1990 conterminous U.S. AVHRR data set. *Photogrammetric Engineering and Remote Sensing*, **58**, 809–813.

Engel, B.A., Srinivasan, R. & Rewerts C. (1993) A spatial decision support system for modeling and managing agricultural non-point-source pollution. In: Goodchild, M.R., Parks, B.O. & Steyaert, L.T. (eds) *Environmental Modeling with GIS*, 231–237. Oxford University Press, New York, NY.

Englund, E. & Sparks, A. (1988) *GEO-EAS (Geostatistical Environmental Assessment Software User's Guide*. Rep. No. EPA/600/4-88/033. U.S. Environmental Protection Agency, Las Vegas, NV.

Environmental Systems Research Institute, Inc. (ESRI) (1992) *Cell-Based Modeling with GRID 6.1. Hydrologic and Distance Modeling Tools*. Supplement. Environmental Systems Research Institute, Redlands, CA.

Escobar, D.E., Bowen, R.L., Gausman, H.W. & Cooper, G.R. (1983) Use of near-infrared video recording system for the detection of freeze-damaged citrus leaves. *Journal of the Rio Grande Valley Horticultural Society*, **36**, 61–66.

Evans, I.S. (1977) The selection of class intervals. *Institute of British Geographers (New Series), Transactions*, **2**, 89–124.

Evans, I.S. (1980) An integrated system of terrain analysis and slope mapping. *Zeitschrift fur Geomorphologie Supplement*, **36**, 274–295.

Everitt, J.H. & Nixon, P.R. (1985) Video imagery: a new remote sensing tool for range management. *Journal of Range Management*, **38**, 421–424.

Everitt, J.H., Escobar, D.E., Gerbermann, A.H. & Alaniz, M.A. (1988) Detecting saline soils with video imagery. *Photogrammetric Engineering and Remote Sensing*, **54**, 1283–1287.

Everitt, J.H., Hussey, M.A., Escobar D.E., Nixon, P.R. & Pinkerton, B. (1986) Assessment of grassland phytomass with airborne video imagery. *Remote Sensing of Environment*, **20**, 299–306.

Fahrig, L. (1991) Simulation methods for developing general landscape-level hypotheses of single-species dynamics. In: Turner, M.G. & Gardner, R.H., *Quantitative Methods in Landscape Ecology*, 415–442. Springer-Verlag, New York, NY.

Fairbanks, D., McGwire, K., Cayocca, K., LeNay, J. & Estes, J. (1996) Sensitivity to climate change of floristic gradients in vegetation communities. In: Goodchild, M.R., Steyaert, L.T., Parks, B.O. *et al.* (eds) *GIS and Environmental Modeling: Progress and Research Issues*, 135–140. GIS World Books, Fort Collins, CO.

Falconer, K (1990) *Fractal Geometry. Mathematical Foundations and Applications.* Wiley, Chichester.

Featherstone, W. & Langley, R.B. (1997) Coordinates and datums and maps! Oh my! *GPS World*, **8**(1), 34–41.

Federal Geographic Data Committee (FGDC) (1992) *Manual of Federal Geographic Data Products.* Vigyan, Falls Church, VA.

Federal Geographic Data Committee (FGDC) (1994) *Content Standards for Digital Geospatial Metadata (June 8).* Federal Geographic Data Committee, Washington, D.C.

Fedra, K. (1993) GIS and environmental modeling. In: Goodchild, M.R., Parks, B.O. & Steyaert, L.T. (eds) *Environmental Modeling with GIS*, 35–50. Oxford University Press, New York, NY.

Fedra, K. & Loucks, D.P. (1985) Interactive computer technology for planning and policy modeling. *Water Resources Research*, **21**(2), 114–122.

Fedra, K, & Reitsma, R.F. (1990) Decision support and geographical information systems. In: Scholten, H.J. & Stillwell, J.C.H. (eds) *Geographical Information Systems for Urban and Regional Planning*, 177–186. Kluwer, Dordrecht.

Fedra, K. & Winkelbauer, L. (1991) MEXSES: an expert system for environmental screening. In: *Proceedings Seventh IEEE Conference on Artificial Intelligence Applications.* IEEE Computer Society Press, Los Alamitos, CA.

Fegeas, R.G., Cascio, J.L. & Lazar, R.A. (1992) An overview of FIPS 173, the Spatial Data Transfer Standard. *Cartography and Geographic Information Systems*, **19**, 278–293.

Fegeas, R.G., Claire, R.W., Guptill, S.C., Anderson, K.E. & Hallam, C.A. (1983) *Land Use and Land Cover Digital Data.* Geological Survey Circular 895-E. U.S. Geological Survey, Washington, D.C.

Ferguson, R.L., Wood, L.L. & Graham, D.B. (1993) Monitoring spatial change in seagrass habitat with aerial photography. *Photogrammetric Engineering and Remote Sensing*, **59**, 1033–1038.

Folorunso, O.A. & Rolston, D.E. (1985) Spatial variability of field-measured denitrification gas fluxes. *Soil Science Society of America Journal*, **48**, 1214–1219.

Foody, G.M. (1992) A fuzzy sets approach to the representation of vegetation continua from remotely sensed data: an example from lowland heath. *Photogrammetric Engineering and Remote Sensing*, **58**, 221–225.

Forman R.T.T. & Godron, M. (1986) *Landscape Ecology.* Wiley, New York, NY.

Fowler, R.J. & Little, J.J. (1979) Automatic extraction of irregular network digital terrain models. *Computer Graphics*, **13**, 199–207.

Fuller, R.M., Groom, G.B. & Jones, A.R. (1994) The Land Cover Map of Great Britain: an automated classification of Landstat Thematic Mapper data. *Photogrammetric Engineering and Remote Sensing*, **60**(5), 553–562.

Gajem, Y.M., Warrick, A.W. & Myers, D.E. (1983) Spatial dependence of physical properties of a Typic Torrifluvent soil. *Soil Science Society of America Journal*, **45**, 709–715.

Gildea, M.P., Moore, B. & Vorosmarty, C.J. (1986) A global model of nutrient cycling: 1. introduction, model structure, and terrestrial mobilization of nutrients. In: Correll, D.L. (ed.) *Watershed Research Perspectives*, 1–31. Smithsonian Institution Press, Washington, D.C.

Gilruth, P.T. & Hutchinson, C.F. (1990) Assessing deforestation in the Guinea highlands of west Africa using remote sensing. *Photogrammetric Engineering and Remote Sensing*, **56**(10), 1375–1382.

GIS World (1996) *GIS World Sourcebook*. GIS World, Fort Collins, CO.

Godron, M. (1966) Application de la théorie de l'information à l'étude de l'homogénéité et de la structure de la végétation. *Oecol, Plantarum*, **1**, 187–197.

Gong, P. & Chen, J. (1992) Boundary uncertainty in digitized maps I: some possible determination methods. In: *GIS/LIS'92 Proceedings, San Jose, CA, Volume 1*, 274–281. American Society for Photogrammetry and Remote Sensing, Bethesda, MD.

Gonzales, R.C. & Woods, R.E. (1992) *Digital Image Processing*. Addison-Wesley, Reading, MA.

Goodchild, M.F. (1985) Geographic information systems in undergraduate geography: a contemporary dilemma. *The Operational Geographer*, **8**, 34–38.

Goodchild, M.F. (1993) Data models and data quality: problems and prospects. In: Goodchild, M.F., Parks, B.O. & Steyaert, L.T. (eds) *Environmental Modeling with GIS*, 94–103. Oxford University Press, New York, NY.

Goodchild, M.F. & Gopal, S. (eds) (1989) *Accuracy of Spatial Databases*. Taylor & Francis, London.

Goodchild, M.F. & Kemp, K.K. (1990) *NCGIA Core Curriculum: Technical Issues in GIS*. National Center for Geographic Information and Analysis, University of California, Santa Barbara, CA.

Goodchild, M.R., Parks, B.O. & Steyaert, L.T. (eds) (1993) *Environmental Modeling with GIS*. Oxford University Press, New York, NY.

Goodchild, M.R., Steyaert, L.T., Parks, B.O. *et al.* (eds) (1996) *GIS and Environmental Modeling: Progress and Research Issues*. GIS World Books, Fort Collins, CO.

Goossens, R., Ongena, T., D'Haluin, E. & Larnoe, G. (1993) The use of remote sensing (SPOT) for the survey of ecological patterns, applied to two different ecosystems in Belgium and Zaire. In: Haines-Young, R., Green, D.R. & Cousins, S.H. (eds) *Landscape Ecology and Geographic Information Systems*, 147–159. Taylor & Francis, New York, NY.

Gosz J.R., Dahm, C.N. & Risser, P.G. (1988) Long-path FTIR measurement of atmospheric trace gas concentrations. *Ecology*, **69**, 1326–1330.

Goward, S.N., Tucker, C.J. & Dye, D.G. (1986) Northern American vegetation patterns observed with meteorological satellite data. In: Dyer, M.I. & Crossley, D.A., Jr. (eds) *Coupling of Ecological Studies with Remote Sensing Potentials at Four Biosphere Reserves in the United States*, 96–115. U.S. Man and the Biosphere Program, Department of State, Washington, D.C.

GPS World (1997) GPS World receiver survey. *GPS World*, **8**(1), 42–60.

Green, D.R., Cummins, R., Wright, R. & Miles, J. (1993) A methodology for acquiring information on vegetation succession from remotely sensed imagery. In: Haines-Young, R., Green, D.R. & Cousins, S.H. (eds) *Landscape Ecology and Geographic Information Systems*, 111–128. Taylor & Francis, London.

Green, P.J. & Sibson, R. (1978) Computing Dirichlet tesselations in the plane. *The Computing Journal*, **21**, 168–173.

Green, R. (1964) *The Storage and Retrieval of Data for Water Quality Control*. Public Health Service Publication No. 1263. U.S. Department of Health, Education, and Welfare, Public Health Service, Washington, D.C.

Greenhood, D. (1964) *Mapping*. University of Chicago Press, Chicago, IL.

Greenlee, D.D. (1987) Raster and vector processing for scanned linework. *Photogrammetric Engineering and Remote Sensing*, **53**, 1383–1387.

Griffiths, G.H., Smith, J.M., Veitch, N. & Aspinall, R. (1993) The ecological interpretation of satellite imagery with special reference to bird habitats. In: Haines-Young, R., Green, D.R. & Cousins, S.H. (eds) *Landscape Ecology and Geographic Information Systems*, 255–272. Taylor & Francis, New York, NY.

Gulinck, H., Walpot, O. & Janssens, P. (1993) Landscape structural analysis of central Belgium using SPOT data. In: Haines-Young, R., Green, D.R. & Cousins, S.H. (eds) *Landscape Ecology and Geographic Information Systems*, 129–139. Taylor & Francis, New York, NY.

Hall, F.G., Botkin, D.B., Strebel, D.E., Woods, K.D. & Goetz, S.J. (1991) Large-scale patterns of forest succession as determined by remote sensing. *Ecology*, **72**, 628–640.

Halsch, P.J. (1992) GPS stakes the highway. *GIS World*, **3**(5), 38.

Hamlett, J.M., Miller, D.A., Day, R.L., Petersen, G.W., Baumer, G.M. & Russo, J. (1992) Statewide GIS-based ranking of watersheds for agricultural pollution prevention. *Journal of Soil and Water Conservation*, **47**, 399–404.

Hardy, E.E. & Shelton, R.L. (1970) Inventorying New York's land use and natural resources. *New York's Food and Life Sciences*, **3**(4), 4–7.

Harris, P.M. & Ventura, S.J. (1995) The integration of geographic data with remotely sensed imagery to improve classification in an urban area. *Photogrammetric Engineering and Remote Sensing*, **61**(8), 993–998.

Harris, P.M., Kim, K.H., Ventura, S.J., Thum, P.G. & Prey, J. (1991) Linking a GIS with an urban nonpoint source pollution model. In: *GIS/LIS '91 Proceedings*, 606–616. American Society for Photogrammetry and Remote Sensing, Bethesda, MD.

Hatfield, J.L., Millard, J.P. & Goettelman, R.G. (1982) Variability of surface temperature in agricultural fields of central California. *Photogrammetric Engineering and Remote Sensing*, **48**, 1319–1325.

Heinen, J. & Cross, G.H. (1983) An approach to measure interspersion, juxtaposition, and spatial diversity from cover-type maps. *Wildlife Society Bulletin*, **11**, 232–237.

Herr, A.M. & Queen, L.P. (1993) Crane habitat evaluation using GIS and remote sensing. *Photogrammetric Engineering and Remote Sensing*, **59**, 1531–1538.

Hession, W.C. & Shanholtz, V.O. (1988) A geographic information system for targeting nonpoint source agricultural pollution. *Journal of Soil and Water Conservation*, **43**, 264–266.

Heuvelink, G.B.M. (1993) Error propagation in quantitative spatial modeling: applications in Geographical Information Systems. Ph.D. thesis, University of Utrecht, The Netherlands.

Hinckley, T.K. & Walker, J.W. (1993) Obtaining and using low-altitude/large-scale imagery. *Photogrammetric Engineering and Remote Sensing*, **59**(3), 310–318.

Hodgson, M.E. (1988) Monitoring wood stork foraging habitat using remote sensing and geographic information systems. *Photogrammetric Engineering and Remote Sensing*, **54**, 1601–1607.

Hsu, Y-M., Myers, G., Clark, T., Jakes, D. & Schlotthauer, J. (1993) Integrating GIS/GPS into Minnesota's statewide ground water monitoring and assessment program. In: *GIS/LIS'93 Proceedings, Minneapolis, MN*, 323–330. American Society for Photogrammetry and Remote Sensing. Bethesda, MD

Hudson, W.D. & Krogulecki, M.J. (1987) A comparison of the land cover and use layer of three statewide geographic information systems in Michigan. In: *Proceedings, GIS'87, San Francisco, CA, Volume 1*, 276–287. American Society for Photogrammetry and Remote Sensing, Falls Church, VA.

Humphreys, R.G. (1992) Monitoring Morecambe Bay. *GPS World*, **3**(9), 27–32.

Hunsaker, C.T. & Levine, D.A. (1995) Hierarchical approaches to the study of water quality in rivers. *BioScience*, **45**, 193–203.

Hunsaker, C.T., Nisbet, R.A., Lam, D.C.L. *et al.* (1993) Spatial models of ecological systems and processes: the role of GIS. In: Goodchild, M.R., Parks, B.O. & Steyaert, L.T. (eds) *Environmental Modeling with GIS*. Oxford University Press, New York, NY.

Hurn, J. (1989) *GPS: A Guide to the Next Utility*. Trimble Navigation, Sunnyvale, CA.

Hutchinson, M. (1993) Development of a continent-wide DEM with applications to terrain and climate analysis. In: Goodchild, M.R., Parks, B.O. & Steyaert, L.T. (eds) *Environmental Modeling with GIS*, 392–399. Oxford University Press, New York, NY.

Hyman, J.B., McAninch, J.B. & DeAngelis, D.L. (1991) An individual-based simulation model of herbivory in a heterogeneous landscape. In: Turner, M.G. & Gardner, R.H. *Quantitative Methods in Landscape Ecology*, 443–475. Springer-Verlag, New York, NY.

Ilbery, B.W. & Evans, N.J. (1989) Estimating land loss on the urban fringe: a comparison of the agricultural census and aerial photograph/map evidence. *Geography*, **74**, 214–221.

Imhoff, M.L. & Gesch, D.B. (1990) Derivation of a sub-canopy digital terrain model of a flooded forest using synthetic aperture radar. *Photogrammetric Engineering and Remote Sensing*, **56**(8), 1155–1162.

Iverson, L.R. & Risser, P.G. (1987) Analyzing long-term changes in vegetation with geographic information systems and remotely sensed data. *Advances in Space Research*, **7**, 183–194.

Jacobs, J. (1974) Quantitative measurement of food selection: a modification of the forage ratio and Ivlev's electivity index. *Oecologia*, **14**, 413–417.

Jakubauskas, M.E. (1990) Assessment of vegetation change in a fire-altered forest landscape. *Photogrammetric Engineering and Remote Sensing*, **56**, 371–377.

Jankowski, P. & Haddock, G. (1996) Integrated nonpoint source pollution modeling system. In: Goodchild, M.R., Steyaert, L.T., Parks, B.O. *et al. (eds) GIS and Environmental Modeling: Progress and Research Issues*, 209–212. GIS WorldBooks, Fort Collins, CO.

Jeffers, J.N.R. (1988) *Practitioner's Handbook on the Modeling of Dynamic Change in Ecosystems*. Wiley, New York, NY.

Jennings, C.A., Vohs, P.A. & Dewey, M.R. (1992) Classification of a wetland area along the Upper Mississippi River with aerial videography. *Wetlands*, **12**(3), 163–170.

Jensen, J.R. (1986) *Introductory Digital Image Progressing: A Remote Sensing Perspective*. Prentice-Hall, Englewood Cliffs, NJ.

Jensen, J.R., Cowen, D.J., Althausen, J.D., Narumalani, S. & Weatherbee, O. (1993) An evaluation of the CoastWatch change detection protocol in South Carolina. *Photogrammetric Engineering and Remote Sensing*, **59**, 1039–1046.

Jensen, S.K. & Domingue, J.O. (1988) Extracting topographic structure from raster elevation data for geographic information system analysis. *Photogrammetric Engineering and Remote Sensing*, **54**, 1593–1600.

Johnson, B.L. & Johnston, C.A. (1995) Relationship of lithology and geomorphology to erosion of the western Lake Superior coast. *Journal of Great Lakes Research*, **21**, 3–16.

Johnson, L.B., Host, G.E., Jordan, J.K. & Rogers, L.L. (1991) Use of GIS for landscape design in natural resource management: habitat assessment and management for the female black bear. In: *GIS/LIS'91 Proceedings*, 507–517. American Society for Photogrammetry and Remote Sensing, Bethesda, MD.

Johnson, W.C. & Sharpe, D.M. (1976) An analysis of forest dynamics in the northern Georgia Piedmont. *Forest Science*, **22**, 307–322.

Johnston, C.A. (1977) Considerations for Wetland Inventories in Wisconsin. M.S. thesis, University of Wisconsin, Madison, WI.

Johnston, C.A. (1989) Ecological research applications of geographic information systems. In: *GIS/LIS'89 Proceedings*, 569–577. American Society for Photogrammetry and Remote Sensing, Bethesda, MD.

Johnston, C.A. (1990) GIS: more than just a pretty face. *Landscape Ecology*, **4**, 3–4.

Johnston, C.A. (1993a) Introduction to quantitative methods and modeling in community, population, and landscape ecology. In: Goodchild, M.R., Parks, B.O & Steyaert, L.T. (eds) *Environmental Modeling with GIS*, 276–283. Oxford University Press, New York, NY.

Johnston, C.A. (1993b) Material fluxes across wetland ecotones in northern landscapes. *Ecological Applications*, **3**, 424–440.

Johnston, C.A. & Bonde, J.P. (1989) Quantitative analysis of ecotones using a geographic information system. *Photogrammetric Engineering and Remote Sensing*, **55**, 1643–1647.

Johnston, C.A. & Naiman, R.J. (1990a) Aquatic patch creation in relation to beaver population trends. *Ecology*, **71**, 1617–1621.

Johnston, C.A. & Naiman, R.J. (1990b) The use of a geographic information system to analyze long-term landscape alteration by beaver. *Landscape Ecology*, **4**, 5–19.

Johnston, C.A. & Salés, J. (1994) GIS for erosion hazard prediction along Lake Superior. *Journal of the Urban and Regional Information Systems Association*, **6**(1), 57–62.

Johnston, C.A., Allen, B., Bonde, J., Salés, J. & Meysembourg, P. (1991) Land use and water resources in the Minnesota North Shore drainage basin. Technical Report *NRRI/TR-94/01*. Natural Resources Research Institute, University of Minnesota, Duluth, MN.

Johnston, C.A., Cohen, Y. & Pastor, J. (1996) Modeling of spatially static and dynamic ecological processes. In: Goodchild, M.R., Steyaert, L.T., Parks, B.O. *et al.* (eds) *GIS and Environmental Modeling: Progress and Research Issues*, 149–154. GIS World Books, Boulder, CO.

Johnston, C.A., Detenbeck, N.E., Bonde, J.P. & Niemi, G.J. (1988) Geographic information systems for cumulative impact assessment. *Photogrammetric Engineering and Remote Sensing*, **54**, 1609–1615.

Johnston, C.A., Detenbeck, N.E. & Niemi, G.J. (1990) The cumulative effect of wetlands on stream water quality and quantity: a landscape approach. *Biogeochemistry*, **10**, 105–141.

Johnston, C.A., Marlett, W. & Riggle, M. (1988) Application of a computer-automated wetlands inventory to regulatory and management problems. *Wetlands*, **8**, 43–52.

Johnston, C.A., Pastor, J. & Naiman, R.J. (1993) Effects of beaver and moose on boreal forest landscapes. In: Cousins, S.H., Haines-Young, R. & Green, D. (eds) *Landscape Ecology and Geographical Information Systems*, 236–254. Taylor & Francis, London.

Johnston, C.A., Pastor, J. & Pinay, G. (1992) Quantitative methods for studying landscape boundaries. In: di Castri, F. & Hansen, A. (eds) *Landscape Boundaries: Consequences for Biotic Diversity and Ecological Flows*, 107–125. Springer-Verlag, New York, NY.

Johnston, C.A., Schubauer-Berigan, J.P. & Bridgham, S.D. (1997) The potential role of riverine wetlands as buffer zones. In: Haycock, N.E., Burt, T.P., Goulding, K.W.T. & Pinay, G. *Buffer Zones: Their Processes and Potential in Water Protection.* Quest Environmental, Harpenden, Herts.

Johnston, C.A., Sersland, C.A., Bonde, J., Pomroy-Petry, D. & Meysembourg, P. (1996) Constructing detailed vegetation databases from field data and airborne videography. *Proceedings of the Third International Conference/Workshop on Integrating GIS and Environmental Modeling.* National Center for Geographic Information and Analysis, Santa Barbara, CA (CD ROM).

Johnston, K.M. (1987) Natural resource modeling in the geographic information system environment. *Photogrammetric Engineering and Remote Sensing*, **53**, 1405–1410.

Johnston, K.M. (1992) Using statistical regression analysis to build three prototype GIS wildlife models. In: *Proceedings, GIS/LIS'92, San Jose, CA, Volume 1*, 374–386. American Society for Photogrammetry and Remote Sensing, Bethesda, MD.

Jones, N.L., Wright, S.G. & Maidment, D.R. (1990) Watershed delineation with triangle-based terrain models. *Journal of Hydraulic Engineering*, **116**, 1232–1251.

Joria, P.E. & Jorgenson, J.C. (1996) Comparison of three methods for mapping tundra with Landsat digital data. *Photogrammetric Engineering and Remote Sensing*, **62**(2), 163–169.

Journel, A.G. & Huijbregts, C.J. (1978) *Mining Geostatistics*. Academic Press, London.

Kandel, A. (1985) *Fuzzy Mathematical Techniques with Applications*. Addison-Wesley, Reading, MA.

Kapetsky, J.M., McGregor, L. & Nanne, H.E. (1987) A geographical information system to assess opportunities for aquaculture development: a FAO – UNEP/GRID study in Costa Rica. In: *Proceedings, GIS '87, San Francisco, CA, Volume 2*, 519–535. American Society for Photogrammetry and Remote Sensing, Falls Church, VA.

Kessell, S.R. (1979) *Gradient Modeling: Resource and Fire Management*. Springer-Verlag, New York, NY.

Kessell, S.R. (1990) An Australian geographical information and modeling system for natural area management. *International Journal of Geographical Information Systems*, **4**, 333–362.

Kessell, S.R. (1996) The integration of empirical modeling, dynamic process modeling, visualization, and GIS for bushfire decision support in Australia. In: Goodchild, M.R., Steyaert, L.T., Parks, B.O. *et al.* (eds) *GIS and Environmental Modeling: Progress and Research Issues*, 367–371. GIS World Books, Fort Collins, CO.

Kessell, S.R. & Beck, J.A. (1991) Development and implementation of forest fire modeling and decision support systems in Australia. In: *Proceedings, GIS/LIS '91, Atlanta, GA, Volume 2*, 805–816. American Society for Photogrammetry and Remote Sensing, Bethesda, MD.

Kim, K. & Ventura, S. (1993) Large-scale modeling of urban nonpoint source pollution using a geographic information system. *Photogrammetric Engineering and Remote Sensing*, **59**, 1539–1544.

Kittel, T.G.F., Ojima, D.S., Schimel, D.S. *et al.* (1996) Model GIS integration and data set development to assess terrestrial ecosystem vulnerability to climate change. In: Goodchild, M.R., Steyaert, L.T., Parks, B.O. *et al* (eds). *GIS and Environmental Modeling: Progress and Research Issues*, 293–298. GIS World Books, Fort Collins, CO.

Klazura, G. & Imy, D.A. (1993) A description of the initial set of analysis products available from the NEXRAD WSR-88D system. *Bulletin of the American Meterological Society*, **74**, 1293–1311.

Kleusberg, A (1992) Precise differential positioning and surveying. *GPS World*, **3**(7), 50–52.

Knowlton, J. (1992) Finding the elusive bass with GPS. *GPS World*, **3**(5), 33.

Kollias, V.J. & Voliotis, A. (1991) Fuzzy reasoning in the development of geographical information systems. *International Journal of Geographical Information Systems*, **5**, 209–223.

Krige, D.G. (1966) Two-dimensional weighted moving average trend surfaces for ore evaluation. In: *Proceedings of the Symposium on Mathematical Statistics and Computer Applications in Ore Valuation*, 13–38. Johannesburg, South Africa.

Krol, E. (1992) *The Whole Internet*. O'Reilley & Associates, Sebastopol, CA.

Krummel, J.R., Dunn, C.P., Eckert, T.C. & Ayers, A.J. (1996) A technology to analyze spatiotemporal landscape dynamics: application to Cadiz Township (Wisconsin). In: Goodchild, M.R., Steyaert, L.T., Parks, B.O. *et al.* (eds) *GIS and Environmental Modeling: Progress and Research Issues*, 169–174. GIS World Books, Fort Collins, CO.

Lakhtakia, M.N., Miller, D.A., White, R.A. & Smith, C.B. (1996) GIS as an integrative tool in climate and hydrology modeling. In: Goodchild, M.R., Steyaert, L.T., Parks, B.O. *et al.* (eds) *GIS and Environmental Modeling: Progress and Research Issues*, 309–312. GIS World Books, Fort Collins, CO.

Lam, D.C.L. & Swayne, D.A. (1991) Integrating database, spreadsheet, graphics, GIS, statistics, simulation models, and expert systems: experiences with the RAISON system on microcomputers. In: Loucks, D.P., de Costa, J.R. (eds) *NATO ASI Series, Volume G 26*, 429–459. Springer-Verlag, Heidelberg.

Lam, D.C.L., Wong, I., Swayne, D.A. & Storey, J. (1992) A knowledge-based approach to regional acidification modeling. *Environmental Monitoring and Assessment*, **23**, 83–97.

Lam, N.S.N. & DeCola, L. (eds) *Fractals in Geography*. PTR Prentice-Hall, Englewood Cliffs, NJ.

Land Management Information Center (1992) *EPPL7 Users Guide, Version 2.1.* Land Management Information Center, St. Paul, MN.

Langley, R.B. (1991) The GPS receiver: an introduction. *GPS World*, **2**(1), 50–53.

Langley, R.B. (1992) Basic geodesy for GPS. *GPS World*, **2**(3), 44–49.

Langley, R.B. (1993) The GPS observables. *GPS World*, **4**(4), 52–59.

Lanter, D.P. (1992) *GEOLINEUS: Data Management and Flowcharting for ARC/INFO.* Technical Software Series S-92-2. National Center for Geographic Information and Analysis, Santa Barbara, CA.

Lanter, D.P. (1994) A lineage metadata approach to removing redundancy and propagating updates in a GIS database. *Cartography and Geographic Information Systems*, **21**(2), 91–98.

Lauver, C.L. & Whistler, J.L. (1992) Using Landsat TM imagery to estimate coverage of natural grassland and rare species habitat. In: *ASPRS/ACSM Technical Papers, Volume 1*, 241–246. American Society for Photogrammetry and Remote Sensing, Bethesda, MD.

Law, A.M. & Kelton, W.D. (1982) *Simulation Modeling and Analysis*. McGraw-Hill, New York, NY.

Leckie, D.G. (1990) Synergism of synthetic aperture radar and visible-infrared data for forest type discrimination. *Photogrammetric Engineering and Remote Sensing*, **56**(9), 1237–1246.

Lee, J.K., Park, R.A., Mausel, P.W. & Howe, R.C. (1991) GIS-related modeling of impacts of sea level rise on coastal areas. In: *Proceedings, GIS/LIS '91. Atlanta, GA, Volume 1*, 356–367. American Society for Photogrammetry and Remote Sensing, Bethesda, MD.

Lees, B (1996) Sampling strategies for machine learning using GIS. In: Goodchild, M.R., Steyaert, L.T., Parks, B.O. *et al.* (eds) *GIS and Environmental Modeling: Progress and Research Issues*, 39–42. GIS World Books, Fort Collins, CO.

Legendre, P. (1993) Spatial autocorrelation: trouble or new paradigm? *Ecology*, **74**, 1659–1673.

Leshkevich, G.A., Schwab, D.J. & Muhr, G.C. (1993) Satellite environmental monitoring of the Great Lakes: a review of NOAA's Great Lakes CoastWatch Program. *Photogrammetric Engineering and Remote Sensing*, **59**(3), 371–379.

Leung, Y., Goodchild, M.F. & Lin, C.C. (1992) Visualization of fuzzy scenes and probability fields. In: *Proceedings, 5th International Symposium on Spatial Data Handling, Volume 2*, 480–490. Humanities and Social Sciences Computing Lab, University of South Carolina, SC.

Levine, D.A., Hunsaker, C.T., Timmins, S.P. & Beauchamp, J.J. (1993) *A Geographic Information System Approach to Modeling Nutrient and Sediment Transport.* Oak Ridge National Laboratory Publication ORNL-6736, Oak Ridge, TN.

Li, H. & Reynolds, J.F. (1993) A new contagion index to quantify spatial patterns of landscapes. *Landscape Ecology*, **8**, 155–162.

Light, D.L. (1993) The national aerial photography program as a geographical information system resource. *Photogrammetric Engineering and Remote Sensing*, **59**(1), 61–65.

Lillesand, TM. & Kiefer, R.W. (1994) *Remote Sensing and Image Interpretation*, 3rd edn. Wiley, New York, NY.

Lo, T.H.C., Ries, T.F. (1992) Change detection of seagrass in Tampa Bay, Florida, using GIS technology. In: *Proceedings, ASPRS/ACSM/RT '92 Convention on Monitoring and*

Mapping Global Change, Volume 3, 240–253. American Society for Photogrammetry and Remote Sensing, Bethesda, MD.

Loveland, T.R. & Scholz, D.K. (1993) Global data set development and data distribution activities at the U.S. Geological Survey's EROS Data Center. In: *ASPRS Technical Papers, Remote Sensing, Looking to the Future with an Eye to the Past*, Vol. II. American Society for Photogrammetry and Remote Sensing, Bethesda, MD.

Loveland, T.R., Merchant, J.W., Ohlen, D.O. & Brown, J.F. (1991) Development of a land-cover characteristics database for the conterminous U.S. *Photogrammetric Engineering and Remote Sensing*, **57**, 1453–1463.

Lowell, K. (1991) Utilizing discriminate function analysis with a geographical information system to model ecological succession spatially. *International Journal of Geographical Information Systems*, **5**, 175–191.

Luczkovich, J.J., Wagner, T.W., Michalek, J.L. & Stoffle, R.W. (1993) Discrimination of coral reefs, seagrass meadows, and sand bottom types from space: a Dominican Republic case study. *Photogrammetric Engineering and Remote Sensing*, **59**(3), 385–389.

Ludwig, J.A. & Cornelius, J.M. (1987) Locating discontinuities along ecological gradients. *Ecology*, **68**, 448–450.

Lupien, A.E., Moreland, W.H. & Dangermond, J. (1987) Network analysis in geographic information systems. *Photogrammetric Engineering and Remote Sensing*, **53**, 1417–1421.

McBratney, A.B. & Webster, R. (1983) How many observations are needed for regional estimation of soil properties? *Soil Science*, **135**, 177–183.

McGarigal, K. & Marks, B.J. (1994) *Fragstats: Spatial Pattern Analysis Program for Quantifying Landscape Structure*. Oregon State University, Corvallis, OR.

McGuire, J., Collins, W.G. & Elgy, J. (1990) Identification and mapping of diminishing natural resources, using Landsat TM and SPOT: a case study. In: *ACSM-ASPRS Technical Papers*, B185–B189. American Society for Photogrammetry and Remote Sensing, Bethesda, MD.

McHarg, I. (1969) *Design with Nature*. Doubleday & Co, Garden City, NJ.

McKenna, P.C. (1992) GPS in the Gobi: dinosaurs among the dunes. *GPS World*, **3**(6), 20–26.

Mackey, B.G., Sims, R.A., Baldwin, K.A. & Moore, I.D. (1996) Spatial analysis of boreal forest ecosystems: results from the Rinker Lake case study. In: Goodchild, M.R., Parks, B.O. & Steyaert, L.T. (eds) *Environmental Modeling with GIS*, 187–190. Oxford University Press, New York, NY.

Maggio, R.C. & Long, D.W. (1991) Developing thematic maps from point sampling using Thiessen polygon analysis. In: *Proceedings, GIS/LIS'91, Atlanta, GA, Volume 1*, 1–10. American Society for Photogrammetry and Remote Sensing, Bethesda, MD.

Maggio, R.C., Wunneburger, D.F. & Long, D.W. (1993) Georeferenced videography and dynamic segmentation for determining housing densities along a pipeline corridor. In: *GIS/LIS '93 Proceedings*, 459–465. American Society for Photogrammetry and Remote Sensing, Bethesda, MD.

Magurran, A.E. (1988) *Ecological Diversity and its Measurement*. Princeton University Press, Princeton, NJ.

Maidment, D.R. (1993) GIS and hydrological modeling. In: Goodchild, M.R., Parks, B.O. & Steyaert, L.T. (eds) *Environmental Modeling with GIS*, 147–167. Oxford University Press, New York, NY.

Maidment, D.R. (1996) Environmental modeling within GIS. In: Goodchild, M.R., Steyaert, L.T., Parks, B.O. *et al.* (eds) *GIS and Environmental Modeling: Progress and Research Issues*, 315–323. GIS World Books, Fort Collins, CO.

Maidment, D.R. & Djokic, D. (1990) *Creating an Expert Geographic Information System: The ARC-Nexpert Interface*. Department of Civil Engineering, University of Texas, Austin, TX.

Malanson, G.P., Armstrong, M.P. & Bennett, D.A. (1996) Fragmented forest response to climatic warming and disturbance. In: Goodchild, M.R., Steyaert, L.T., Parks, B.O. *et al.* (eds) *GIS and Environmental Modeling: Progress and Research Issues*, 243–248. GIS World Books, Fort Collins, CO.

Maling, D.H. (1989) *Measurements from Maps: Principles and Methods of Cartometry.* Pergamon Press, Oxford.

Mandelbrot, B.B. (1967) How long is the coast of Britain? Statistical self-similarity and fractional dimension. *Science*, **155**, 636–638.

Mandelbrot, B.B. (1983) *The Fractal Geometry of Nature.* W.H. Freeman & Company, New York, NY.

Mandelbrot, B. & Wallis, J.R. (1969) Some long-run properties of geophysical records. *Water Resources Research*, **5**, 321–340.

Manzer, F.E. & Cooper, G.R. (1982) Use of portable videotaping for aerial infrared detection of potato diseases. *Plant Disease*, **66**, 665–667.

Margalef, R. (1958) Information theory in ecology. *General Systems*, **3**, 36–71.

Marks, D., Dozier, J. & Frew, J. (1984) Automated basin delineation from digital elevation data. *Geo-Processing*, **2**, 299–311.

Matheron, G. (1965) *Les Variables (Régionalisées et Leur Estimation.* Masson, Paris.

Matheron, G. (1971) *The Theory of Regionalized Variables and its Applications.* Les Cahiers du centre de morphologie mathematique de Fontainebleu. Ecole Nationale Superieure des Mines, Paris, France.

Matson, P.A. & Harriss, R.C. (1988) Prospects for aircraft-based gas exchange measurements in ecosystem studies. *Ecology*, **69**, 1318–1325.

Mausel, P.W., Karaska, M.A., Mao, C.Y., Escobar, D.E. & Everitt, J.H. (1989) Insights into water turbidity through computer analysis of multispectral video data. In: *Proceedings of the 12th Biennial Workshop on Color Photography and Videography in the Plant Sciences and Related Fields*, 216–226. American Society for Photogrammetry and Remote Sensing, Bethesda, MD.

Mead, R.A., Cockerham, L.S. & Robinson, C.M. (1988) Mapping gopher tortoise habitat on the Ocala National Forest using a GIS. In: *Proceedings, GIS/LIS '88, San Antonio, TX, Volume 1*, 395–400. American Society for Photogrammetry and Remote Sensing, Falls Church, VA.

Mead, R.A., Sharik, T.L., Prisley, S.P. & Heinen, J.T. (1981) A computerized spatial analysis system for assessing wildlife habitat from vegetation maps. *Canadian Journal of Remote Sensing*, **7**, 34–40.

Merritt, R.W. & Cummins, K.W. (eds) (1984) *An Introduction to the Aquatic Insects of North America*, 2nd edn. Kendal/Hunt Publishing, Dubuque, IA.

Michener, W.K. & Houhoulis, P.F. (1996) Identification and assessment of natural disturbance in forested ecosystems: the role of GIS and remote sensing. *Proceedings of the Third International Conference/Workshop on Integrating GIS and Environmental Modeling.* National Center for Geographic Information and Analysis, Santa Barbara, CA. (CD ROM.)

Michener, W.K., Brunt, J.W. & Stafford, S.G. (eds) (1994) *Environmental Information Management and Analysis: Ecosystem to Global Scales.* Taylor & Francis, London.

Michener, W.K., Miller, A.B. & Nottrott, R. (1990) *Long-Term Ecological Research Network: Core Data Set Catalogue.* Belle W. Baruch Institute for Marine Biology and Coastal Research, University of South Carolina, Columbia, SC.

Michener, W.K., Jefferson, W.H., Karinshak D.A. & Edwards, D. (1992) An integrated geographic information system, global positioning system, and spatio-statistical approach for analyzing ecological patterns at landscape scales. In: *Proceedings, GIS/LIS'92, San Jose, CA, Volume 2*, 564–576. American Society for Photogrammetry and Remote Sensing, Bethesda, MD.

Miller, D.R. (1996) Landscape visualization using DEM data derived from digital photo-

grammetry. *Proceedings of the Third International Conference/Workshop on Integrating GIS and Environmental Modeling*. National Center for Geographic Information and Analysis, Santa Barbara, CA. (CD ROM.)
Miller, K.V. & Conroy, M.J. (1990) Spot satellite imagery for mapping Kirtlands warbler wintering habitat in the Bahamas. *Wildlife Society Bulletin*, **18**, 252–257.
Miller, W.F. (1991) Ground truthing with GPS in Guatemala. *GPS World*, **2**(8), 36–41.
Milne, B.T. (1991) Lessons from applying fractal models to landscape patterns. In: Turner, M.G & Gardner, R.H. (eds) *Quantitative Methods in Landscape Ecology*, 199–235. Springer-Verlag, New York, NY.
Milne, B.T., Johnston, K.M. & Forman, R.T.T. (1989) Scale-dependent proximity of wildlife habitat in a spatially-neutral Bayesian model. *Landscape Ecology*, **2**, 101–110.
Mitasova, H., Mitas, L., Brown, W.M., Gerdes, D.P., Kosinousky, I. & Baker, T. (1996) Modeling spatial and temporal distributed phenomena: new methods and tools for open GIS. In: Goodchild, M.R., Steyaert, L.T., Parks, B.O. *et al.* (eds) *GIS and Environmental Modeling: Progress and Research Issues*, 345–351. GIS World Books, Fort Collins, CO.
Moen, R., Pastor, J., Cohen, Y. & Schwartz, C.C. (1996) Effects of moose movement and habitat use on GPS collar performance. *Journal of Wildlife Management*, **63**, 659–668.
Mohler, R.R.J., Wells, G.L., Hallum, C.R. & Trenchard, M.H. (1986) Monitoring vegetation of drought environments. *BioScience*, **36**, 478–483.
Mooneyhan, D.W. (1988) Applications of geographical information systems within the United Nations Environment Programme. In: Mounsey, H. & Tomlinson, R.F. (eds) *Building Databases for Global Science*, 315–329. Taylor & Francis, London.
Moore, I.D. (1996) Hydrologic modeling and GIS. In: Goodchild, M.R., Steyaert, L.T., Parks, B.O. *et al.* (eds) *GIS and Environmental Modeling: Progress and Research Issues*. 143–148. GIS World Books, Fort Collins, CO.
Moore, I.D., Gessler, P.E., Nielson, G.A. & Peterson, G.A. (1993) Soil attribute prediction using terrain analysis. *Soil Science Society of America Journal*, **57**, 443–452.
Moore, I.D., Turner, A.K., Wilson, J.P. Jensen, S.K. & Band, L.E. (1993) GIS and land-surface–subsurface modeling. In: Goodchild, M.R., Parks, B.O. & Steyaert, L.T. (eds) *Environmental Modeling with GIS*, 196–230. Oxford University Press, New York, NY.
Morehouse, S. & Broekhuysen, M. (1981) *ODESSEY User's Manual*. Laboratory for Computer Graphics, Harvard University, Cambridge, MA.
Moreno, D.D & Heyerdahl, L.A. (1990) Advanced GIS modeling techniques in environmental impact assessment, In: *Proceedings, GIS/LIS '90, Anaheim, CA, Volume 1*, 345–356. American Society for Photogrammetry and Remote Sensing, Bethesda, MD.
Morrison, J.L. & Wortman, K. (1992) Implementing the Spatial Data Transfer Standard. *Cartography and Geographic Information Systems*, **19**, Special Issue.
Musick, H.B. & Grover, H.D. (1991) Image textural measures as indices of landscape pattern. In: Turner, M.G. & Gardner, R.H. (eds) *Quantitative Methods in Landscape Ecology*, 77–103. Springer Verlag, New York, NY.
Nachman, G. (1987) Systems analysis of acarine predator–prey interactions: I. a stochastic simulation of spatial processes. *Journal of Animal Ecology*, **56**, 247–265.
National Academy of Science (1995) *Wetlands: Characteristics and Boundaries*. National Academy of Science Press, Washington, DC.
National Institute of Standards and Technology (1992) *Spatial Data Transfer Standard. Federal Information Processing Standard (FIPS) 173*. National Institute of Standards and Technology, Gaithersburg, MD.
Naveh, Z. (1982) Landscape ecology as an emerging branch of human ecosystem science. *Advances in Ecological Research*, **12**, 189–237.
Nawrocki, T., Johnston, C. & Salés, J. (1994) GIS and modeling in ecological studies: analysis of beaver pond impacts on runoff and its quality. *Technical Report NRRI/TR-94/01*. Natural Resources Research Institute, University of Minnesota, Duluth, MN.

Nazarenko, D., Staples, G. & Aspden, C. (1996) RADARSAT: first images. *Photogrammetric Engineering and Remote Sensing*, **62**, 143–146.

NCGIA (1996) *Proceedings, Third International Conference/Workshop on Integrating GIS and Environmental Modeling, Santa Fe, NM.* National Center for Geographic Information and Analysis, Santa Barbara, CA.

Nellis, M.D. & Briggs, J.M. (1989) The effect of spatial scale on Konza landscape classification using textural analysis. *Landscape Ecology*, **2**, 93–100.

Nemani, R., Running, S.W., Band, L.E. & Peterson, D.L. (1993) Regional hydroecological simulation system: an illustration of the integration of ecosystem models in a GIS. In: Goodchild, M.R., Parks, B.O. & Steyaert, L.T. (eds) *Environmental Modeling with GIS*, 296–304. Oxford University Press, New York, NY.

Nesbit, R.A. & Botkin, D.B. (1993) Integrating a forest growth model with a geographic information system. In: Goodchild, M.R., Parks, B.O. & Steyaert, L.T. (eds) *Environmental Modeling with GIS*, 265–275. Oxford University Press, New York, NY.

Nixon, P.R., Escobar, D.E. & Menges, R.M. (1985) A multiband video system for quick assessment of vegetal condition and discrimination of plant species. *Remote Sensing of Environment*, **17**, 203–208.

Nwadialo, B.E. & Hole, F.D. (1988) A statistical procedure for partitioning soil transects. *Soil Science*, **145**, 58–62.

O'Callaghan, J.F. & Mark, D.M. (1984) The extraction of drainage networks from digital elevation data. *Computer Vision, Graphics and Image Processing*, **28**, 323–344.

Odum, E.P. (1971) Fundamentals of Ecology (3rd edn). W.B. Saunders, Philadelphia, PA.

O'Neill, R.V., Krummel, J.R., Gardner, R.H. *et al.* (1988) Indices of landscape pattern. *Landscape Ecology*, **1**, 153–162.

Ord, J.K. (1979) Time-series and spatial patterns in ecology. In: Cormack, R.M. & Ord, J.K. (eds) *Spatial and Temporal Analysis in Ecology*, 1–94. Statistical Ecology Series Volume 8, International Cooperative Publishing House, Fairland, MD.

Ormsby, J.P. & Lunetta, R.S. (1987) Whitetail deer food availability maps from Thematic Mapper data. *Photogrammetric Engineering and Remote Sensing*, **53**, 1081–1085.

Osborne, L.L. & Wiley, M.J. (1988) Empirical relationships between land use/cover and stream water quality in an agricultural watershed. *Journal of Environmental Management*, **26**, 9–27.

Parent, P. & Church, R. (1989) Evolution of geographic information systems as decision making tools. In: Ripple, W.J. (ed). *Fundamentals of Geographic Information Systems: A Compendium*, 9–18. American Society for Photogrammetry and Remote Sensing, Bethesda, MD.

Parton, W.J. & Risser, P.G. (1980) Impact of management practices on the tallgrass prairie. *Oecologia (Berlin)*, **46**, 223–234.

Pastor, J. & Broschart, M.R. (1990) The spatial pattern of a northern conifer–hardwood landscape. *Landscape Ecology*, **4**, 55–68.

Pastor, J. & Johnston, C.A. (1992) Using simulation models and geographic information systems to integrate ecosystem and landscape ecology. In: Naiman, R.J. (ed.) *Watershed Management: Balancing Sustainability and Environmental Change*, 324–346. Springer-Verlag, New York, NY.

Pastor, J. & Post, W.M. (1988) Response of northern forests to CO_2-induced climate change. *Nature*, **344**, 55–58.

Pastor, J., Bonde, J., Johnston, C.A. & Naiman, R.J. (1993) Markovian analysis of the spatially dependent dynamics of beaver ponds. In: Gardner, R.H. (ed.) *Predicting Spatial Effects in Ecological Systems. Lectures on Mathematics in the Life Sciences, Volume 23*, 5–27. American Mathematical Society, Providence, RI.

Pearlstine, L., McKellar H. & Kitchens, W. (1985) Modeling the impacts of a river diversion on bottomland forest communites in the Santee River Floodplain, South Carolina. *Ecological Modeling*, **29**, 283–302.

Pereira, J.M.C. & Itami, R.M. (1991) GIS-based habitat modeling using logistic multiple regression: a study of the Mt. Graham red squirrel. *Photogrammetric Engineering and Remote Sensing*, **57**, 1475–1486.

Peterson, C. (1990) Into the woods with GPS. *GPS World*, **1**(6), 31–36.

Peucker, T.K. & Chrisman, N. (1975) Cartographic data structures. *The American Cartographer*, **2**, 55–69.

Peucker, T.K., Fowler, R.J., Little, J.J. & Mark, D.M. (1978) The triangulated irregular network. In: *Proceedings, Digital Terrain Models (DTM) Symposium*, 516–540. American Society of Photogrammetry, St. Louis, MO.

Peuquet, D.J. (1984) A conceptual framework and comparison of spatial data models. *Cartographica*, **21**, 66–113.

Philipson, W.R. (ed.) (1996) *The Manual of Photographic Interpretation*, 2nd edn. American Society for Photogrammetry and Remote Sensing, Bethesda, MD.

Pickover, C.A. (1990) *Computers, Pattern, Chaos and Beauty, Graphics from an Unseen World*. St. Martin's Press, New York, NY.

Pielke, R.A., Baron, J., Chase, T. *et al.* (1996) Use of mesoscale models for simulation of seasonal weather and climate change for the Rocky Mountain states. In: Goodchild, M.R., Steyaert, L.T., Parks, B.O. *et al.* (eds) *GIS and Environmental Modeling: Progress and Research Issues*, 99–103. GIS World Books, Fort Collins, CO.

Pielou, E.C. (1959) The use of point-to-plant distances in the study of the pattern of plant populations. *Journal of Ecology*, **47**, 607–613.

Pielou, E.C. (1960) A single mechanism to account for regular, random, and aggregated populations. *Journal of Ecology*, **48**, 575–584.

Pielou, E.C. (1966) Species-diversity and pattern-diversity in the study of ecological succession. *Journal of Theoretical Biology*, **10**, 370–383.

Pielou, E.C. (1984) *The Interpretation of Ecological Data*. Wiley, New York, NY.

Pike, R.J., Thelin, G.P. & Acevedo, W. (1987) A topographic base for GIS from automated TINs and image-processed DEMs. In: *GIS/LIS '87 Proceedings*, 340–351. American Society for Photogrammetry and Remote Sensing, Falls Church, VA.

Poiani, K.A. & Bedford, B.L. (1995) GIS-based nonpoint source pollution modeling: considerations for wetlands. *Journal of Soil and Water Conservation*, **50**(6), 613–619.

Polzer, P.L. Hartzell, B.J., Wynne, R.H., Harris,P.M. & MacKenzie, M.D. (1991) Linking GIS with predictive models: case study in a southern Wisconsin oak forest. In: *Proceedings, GIS/LIS '91, Atlanta, GA, Volume 1*, 49–59. American Society for Photogrammetry and Remote Sensing, Bethesda, MD.

Remillard, M.M., Gruendling, G.K. & Bogucki, D.J. (1987) Disturbance by beaver (*Castor canadensis* Kuhl) and increased landscape heterogeneity. In: Turner, M.G. (ed.) *Landscape Heterogeneity and Disturbance*, 103–123. Springer Verlag, New York, NY.

Richardson, A.J., Menges, R.M. & Nixon, P.R. (1985) Distinguishing weed from crop plants using video sensing. *Photogrammetric Engineering and Remote Sensing*, **51**, 1785–1790.

Riggan, P.J., Brass, J.A. & Lockwood, R.N. (1993) Assessing fire emissions from tropical savanna and forests of central Brazil. *Photogrammetric Engineering and Remote Sensing*, **59**(6), 1009–1015.

Riitters, K.H., O'Neill, R.V., Hunsaker, C.T. *et al.* (1995) A factor analysis of landscape pattern and structure metrics. *Landscape Ecology*, **10**, 23–39.

Ripley, B. (1981) *Spatial Statistics*. Wiley, New York, NY.

Ripple, W.J. (1994) Determining coniferous forest cover and forest fragmentation with NOAA-9 Advanced Very High Resolution Radiometer data. *Photogrammetric Engineering and Remote Sensing*, **60**(5), 533–540.

Ripple, W.J. & Ulshoefer, V.S. (1987) Expert systems and spatial data models for efficient geographic data handling. *Photogrammetric Engineering and Remote Sensing*, **53**, 1431–1433.

Risser, P.G. (1995) The status of the science examining ecotones. *BioScience*, **45**, 318–325.

Roberts, D.W. (1989) Analysis of forest succession with fuzzy graph theory. *Ecological Modeling*, **45**, 261–274.
Robertson, G.P. (1987) Geostatistics in ecology: interpolating with known variance. *Ecology*, **68**, 744–748.
Robertson, G.P., Huston, M.A., Evans, F.C. & Tiedje, J.M. (1988) Spatial variability in a successional plant community: patterns of nitrogen availability. *Ecology*, **69**, 1517–1524.
Robinson, V.B. & Frank, A.U. (1987) Expert systems for geographic information systems. *Photogrammetric Engineering and Remote Sensing*, **53**, 1435–1441.
Rock, B.N., Vogelmann, J.E., Williams, D.L., Vogelmann, A.F. & Hoshizaki, T. (1986) Remote detection of forest damage. *BioScience*, **36**, 439–445.
Roller, N.E.G. & Colwell, J.E. (1986) Coarse-resolution satellite data for ecological surveys. *BioScience*, **36**, 468–475.
Rossi, R.E., Mulla, D.J., Journel, A.G. & Franz, E.H. (1992) Geostatistical tools for modeling and interpreting ecological spatial dependence. *Ecological Monographs*, **62**, 277–314.
Rothermel, R.C. (1972) *A Mathematical Model for Predicting Fire Spread Rate and Intensity in Wildland Fuels*. USDA Forest Service Research Paper INT-115. U.S. Department of Agriculture, Washington, D.C.
Running, S.W. & Thornton, P.E. (1996) Generating daily surfaces of temperature and precipitation over complex topography. In: Goodchild, M.R., Steyaert, L.T., Parks, B.O. *et al.* (eds) *GIS and Environmental Modeling: Progress and Research Issues*, 93–98. GIS World Books, Fort Collins, CO.
Running, S.W., Nemani, R.R., Peterson, D.L. *et al.* (1989) Mapping regional forest evapotranspiration and photosynthesis by coupling satellite data with ecosystem simulation. *Ecology*, **70**, 1090–1101.
Russo, D. (1984) Design of an optimal sampling network for estimating the variogram. *Soil Science Society of America Journal*, **48**, 708–716.
Russo, D. & Bresler, E. (1981) Soil hydraulic properties as stochastic processes: I. An analysis of field spatial variability. *Soil Science Society of America Journal*, **45**, 682–687.
Rutchey, K. & Vilcheck, L. (1994) Development of an Everglades vegetation map using a SPOT image and the global positioning system. *Photogrammetric Engineering and Remote Sensing*, **60**(6), 767–775.
SAS Institute, (1990) *SAS/GRAPH Software. Reference, Version 6*. SAS Institute, Cary, NC.
Sayn-Wittgenstein, L. (1961) Recognition of tree species on air photographs by crown characteristics. *Photogrammetric Engineering*, **27**(5), 792–809.
Scarpace, F.L., Quirk, B.K., Kiefer, R.W. & Wynn, S.L. (1981) Wetland mapping from digitized aerial photography. *Photogrammetric Engineering and Remote Sensing*, **47**, 829–838.
Scepan, J., Davis, F. & Blum, L.L. (1987) A geographic information system for managing California condor habitat. In: *Proceedings, GIS '87, San Francisco, CA, Volume 2*, 476–486. American Society for Photogrammetry and Remote Sensing, Falls Church, VA.
Schloss, J.A. & Rubin, F.A. (1992) A "bottom-up" approach to GIS watershed analysis. In: *GIS/LIS '92 Proceedings*, 672–679. American Society for Photogrammetry and Remote Sensing, Bethesda, MD.
Schreve, R.L. (1966) Statistical law of stream number. *Journal of Geology*, **74**, 17–37.
Schwartz, K.P. & Sideris, M.G. (1993) Heights and GPS. *GPS World*, **4**(2), 50–56.
Scott, J.M. *et al.* (1993) GAP Analysis: a geographic approach to protection of biological diversity. *Wildlife Monographs* **123**, 1–41.
Sersland, C.A., Johnston, C.A. & Bonde, J. (1995) Assessing wetland vegetation with GPS-linked color video image mosaics. In: *Proceedings, 15th Biennial Workshop on Color Photography and Videography in Resource Assessment*, 53–62. American Society for Photogrammetry and Remote Sensing, Bethesda, MD.

Shannon, C.E. & Weaver, W. (1949) *The Mathematical Theory of Communications*. University of Illinois Press, Urbana, IL.
Shaw, D.M. & Atkinson, S.F. (1988) GIS applications for golden-cheeked warbler habitat description. In: *Proceedings, GIS/LIS '88, San Antonio, TX, Volume 1*, 401–406. American Society for Photogrammetry and Remote Sensing, Falls Church, VA.
Sheehan, D.E. (1979) A discussion of the SYMAP program. In: *Harvard Library of Computer Graphics, Mapping Collection Vol. 2: Mapping Software and Cartographic Databases*. 167–179. Laboratory of Computer Graphics, Harvard University, Cambridge. MA.
Shih, S.F. & Doolittle, J.A. (1984) Using radar to investigate organic soil thickness in the Florida Everglades. *Soil Science Society of America Journal*, **48**, 651–656.
Shugart, H.H., Jr., Crow, T.R. & Hett, J.M. (1973) Forest succession models: a rationale and methodology for modeling forest succession over large regions. *Forest Science*, **19**, 203–212.
Shugart, H.H., Jr., Bonan, G.B., Urban, D.L. *et al.* (1991) Computer models and long-term ecological research. In: Risser, P.G. (ed.) *Long-Term Ecological Research: An International Perspective*, 211–239. SCOPE 47, Wiley, Chichester, UK.
Shugart, H.H., Jr. & West, D.C. (1980) Forest succession models. *BioScience*, **30**, 308–313.
Sidle, J.G. & Ziewitz, J.W. (1990) Use of aerial videography in wildlife habitat studies. *Wildlife Society Bulletin*, **18**, 56–62.
Sieg, G.E., Scrivani, J.A. & Smith, J.L. (1987) Incorporating GIS topographic information in forest inventory estimates. In: *GIS/LIS '87 Proceedings*, 423–430. American Society for Photogrammetry and Remote Sensing, Falls Church, VA.
Silveira, J.E., Richardson, J.R. & Kitchens, W.M. (1992) A GIS model of fire, succession and landscape pattern in the Everglades. In: *Proceedings, ASPRS/ACSM/RT '92 Convention on Monitoring and Mapping Global Change, Volume 5*, 89–97. American Society for Photogrammetry and Remote Sensing, Bethesda, MD.
Sinton, D.F. (1977) *The User's Guide to IMGRID: An Information System for Grid Cell Data Structures*. Department of Landscape Architecture, Harvard University, Cambridge, MA.
Sinton, D.F. & Steinitz, C.F. (1969) *GRID: A User's Manual*. Laboratory for Computer Graphics and Spatial Analysis, Harvard University, Cambridge, MA.
Sisson, J.B. & Wierenga, P.J. (1981) Spatial variability of steady-state infiltration rates as a stochastic process. *Soil Science Society of America Journal*, **45**, 699–704.
Sklar, F.H. & Coastanza, R. (1991) The development of dynamic spatial models for landscape ecology: a review and prognosis. In: Turner, M.G. & Gardner, R.H., *Quantitative Methods in Landscape Ecology*, 239–288. Springer-Verlag, New York, NY.
Smith, J.M., Miller, D.R. & Morrice, J.G. (1989) An evaluation of a low-level resolution DTM for use with satellite imagery for environmental mapping and analysis. In: *Remote Sensing for Operational Applications: Proceedings of the 15th Annual Conference on the Remote Sensing Society*, 393–398. Remote Sensing Society, Bristol.
Smith, M.B. & Vidmar, A. (1994) Data set derivation for GIS-based urban hydrological modeling. *Photogrammetric Engineering and Remote Sensing*, **60**, 67–76.
Smith, T.M. & Urban, D.L. (1988) Scale and resolution of forest structural pattern. *Vegetatio* **74**, 143–150.
Soil Survey Staff (1966) Aerial-photo interpretation in classifying and mapping soils. *Agriculture Handbook 294*. U.S. Government Printing Office, Washington, D.C.
Soil Survey Staff (1993) *Soil Survey Manual. USDA Handbook 18*. U.S. Government Printing Office, Washington, DC.
Sokal, R. & Oden, N.L. (1978) Spatial autocorrelation in biology: 1. Methodology. *Biological Journal of the Linnean Society*, **10**, 199–228.
Sombroek, W.G. & Colenbrander, H.J. (1990) Importance of global data in monitoring the soil and water resources. In: Cini, F.G. (ed.) *Global Natural Resource Monitoring and*

Assessments: Preparing for the 21st Century, 60–74. American Society for Photogrammetry and Remote Sensing, Bethesda, MD.

Srinivasan, R. & Arnold, J.G. (1994) Integration of a basin-scale water quality model with GIS. *Water Resources Bulletin*, **30**, 453–462.

Stanley, H.E (1986) Form: an introduction to self-similarity and fractal behavior. In: Stanley, H.E. & Ostrowski, N. (eds) *On Growth and Form: Fractal and Non-Fractal Patterns in Physics*, 21–53. Martinus Nijhoff, Boston, MA.

Star, J. & Estes, J. (1990) *Geographic Information Systems: An Introduction*. Prentice-Hall, Englewood Cliffs, NJ.

Statistical Sciences (1991) *S-Plus User's Manual, Version 3.0*. Statistical Sciences, Inc., Seattle, WA.

Steinitz, C.F., Parker, P. & Jordan, L. (1976) Hand drawn overlays: their history and prospective uses. *Landscape Architecture*, **66**, 444–455.

Stenback, J.M., Travlos, C.B., Barrett, R.G. & Congalton, R.G. (1987) Application of remotely sensed digital data and a GIS in evaluating deer habitat suitability on the Tehama Deer Winter Range. In: *Proceedings, GIS '87, San Francisco, CA, Volume 2*, 440–445. American Society for Photogrammetry and Remote Sensing, Falls Church, VA.

Steube, M.M. & Johnston, D.M. (1990) Runoff volume estimation using GIS techniques. *Water Resources Bulletin*, **26**, 611–620.

Steyaert, L.T. (1993) A perspective on the state of environmental simulation modeling. In: Goodchild, M.R., Parks, B.O. & Steyaert, L.T. (eds) *Environmental Modeling with GIS*, 16–30. Oxford University Pres, New York, NY.

Steyaert, L.T. (1996) Status of land data for environmental modeling and challenges for geographical information systems in land characterization. In: Goodchild, M.R., Steyaert, L.T., Parks, B.O. *et al.* (eds) *GIS and Environmental Modeling: Progress and Research Issues*, GIS World Books, Fort Collins, CO.

Steyaert, L.T. & Goodchild, M.R. (1994) Integrating geographic information systems and environmental simulation models: a status review. In: Michener, W.K., Brunt, J.W. & Stafford, S.G. (eds) *Environmental Information Management and Analysis: Ecosystem to Global Scales*, 333–355. Taylor & Francis, London.

Stoms, D.M., Davis, F.W. & Cogan, C.B. (1992) Sensitivity of wildlife habitat models to uncertainties in GIS data. *Photogrammetric Engineering and Remote Sensing*, **58**, 843–850.

Stone, T.A., Schlesinger, P., Houghton, R.A. & Woodwell, G.M. (1994) Map of the vegetation of South America based on satellite imagery. *Photogrammetric Engineering and Remote Sensing*, **60**, 541–551.

Strahler, A.N. (1957) Quantitative analysis of watershed geomorphology. *Transactions of the American Geophysical Union*, **8**, 913–920.

Sui, Dian-zhi. (1990) A fuzzy mathematical modeling technique and its implementation in geographic information systems: toward a fuzzy cartographic modeling technique. In: *Proceedings, GIS/LIS'90, Anaheim, CA, Volume 1*, 234–243. American Society for Photogrammetry and Remote Sensing, Bethesda, MD.

Thapa, K. & Bossler, J. (1992) Accuracy of spatial data used in geographic information systems. *Photogrammetric Engineering and Remote Sensing*, **6**, 835–841.

Thoen, B (1994) Access the electronic highway for a world of data. *GIS World*, 7(2), 46–49.

Thomasson, J.A., Bennett, C.W., Jackson, B.D. & Mailander, M.P. (1994) Differentiating bottomland tree species with multispectral videography. *Photogrammetric Engineering and Remote Sensing*, **60**(1) 55–59.

Tiner, R.W., Jr. (1990) Use of high-altitude aerial photography for inventory forested wetlands in the United States. *Forest Ecology and Management* **33/34**, 593–604.

Tobler, W. (1988) Resolution, resampling, and all that. In: Mounsey, H. & Tomlin, R. (eds.) *Building Databases for Global Science*, 129–137. Taylor & Francis, London.

Tolkki, S. & Koskelo, I. (1993) Seeing the trees through the forest: GPS streamlines Finnish logging operation. *GPS World*, **4**(10), 20–27.

Tomlin, C.D. (1986) *The IBM Personal Version of the Map Analysis Package.* Laboratory for Computer Graphics and Spatial Analysis, Harvard University, Cambridge, MA.

Tomlin, C.D. (1990) *Geographic Information Systems and Cartographic Modeling.* Prentice-Hall, Englewood Cliffs, NJ.

Tomlin, C.D., Berwick, S.H. & Tomlin, S.M. (1987) The use of computer graphics in deer habitat evaluation. In: Ripple, W.J. (ed.) *Geographic Information Systems for Resource Mangement: A Compendium*, 212–218. American Society for Photogrammetry and Remote Sensing, Falls Church, VA.

Tomlinson, R.F., Calkin, H.W. & Marble, D.F. (1976) *Computer Handling of Geographic Data.* UNESCO, Geneva.

Trangmar, B.B., Yost, R.S., Wade, M.K., Uehara, G. & Sudjadi, M. (1987) Spatial variation of soil properties and rice yield on recently cleared land. *Soil Science Society of America Journal*, **51**, 668–674.

Trotter, C.M. (1991) Remotely-sensed data as an information source for geographical information systems in natural resource management: a review. *International Journal of Geographic Information Systems*, **5**, 225–239.

Tucker, C.J., Holben, B.N. & Goff, T.E. (1984) Intensive forest clearing in Rondonia, Brazil as detected by satellite remote sensing. *Remote Sensing the Environment*, **15**, 255–261.

Tucker, C.J., Townshend, J.R.G. & Goff, T.E. (1985) African land-cover classification using satellite data. *Science*, **227**, 369–375.

Turner, M.G. (1987a) *Landscape Heterogeneity and Disturbance.* Springer Verlag, New York.

Turner, M.G. (1987b) Spatial simulation of landscape changes in Georgia: a comparison of 3 transition models. *Landscape Ecology*, **1**, 29–36.

Turner, M.G. & Dale, V.H. (1991) Modeling landscape disturbance. In: Turner, M.G. & Gardner, R.H. (eds) *Quantitative Methods in Landscape Ecology*, 323–351. Springer-Verlag, New York, NY.

Turner, S.J., O'Neill, R.V., Conley, W., Conley, M.R. & Humphries, H.C. (1991) Pattern and scale: statistics for landscape ecology. In: Turner, M.G. & Gardner, R.H. (eds) *Quantitative Methods in Landscape Ecology*, 17–49. Springer Verlag, New York, NY.

Ulliman, J.J. (1992) Wetland change detection using historical aerial photography. In: *Proceedings, ASPRS/ACSM Annual Convention, Albuquerque, NM, Volume 1*, 480–488. American Society for Photogrammetry and Remote Sensing, Bethesda, MD.

United States Geological Survey (USGS) (1987) *Digital Elevation Models, Data Users Guide 5, National Mapping Program, Technical Instructions.* U.S. Geological Survey, U.S. Department of the Interior, Reston, VA.

Urban, D.L. (1990) A versatile model to simulate forest pattern. *A User's Guide to ZELIG Version 1.0.* Department of Environmental Sciences, University of Virginia, Charlottesville, VA.

U.S. Army Corps of Engineers (1987) *Corps of Engineers Wetlands Delineation Manual.* Department of the Army, Washington, D.C.

U.S. Department of Agriculture (USDA) (1978) Forester's guide to aerial photo interpretation. *Agriculture Handbook No. 308.* U.S. Government Printing Office, Washington, D.C.

U.S. Environmental Protection Agency (US EPA) (1993) *North American Landscape Characterization (NALC). Research Brief EPA/600/S-93/0005.* Environmental Monitoring Systems Laboratory, Las Vegas, NV.

Van Driel, N. & Loveland, T. (1996) The U.S. Geological Survey's land cover characterization program. *Proceedings of the Third International Conference/Workshop on Integrating GIS and Environmental Modeling.* National Center for Geographic Information and Analysis. Santa Barbara, CA. (CD ROM.)

Vande Castle, J.R. (1991) Remote sensing and modeling activities for long-term ecological research. In: *GIS/LIS '91 Proceedings, Volume 2*, 544–550. American Society for Photogrammetry and Remote Sensing, Bethesda, MD.

Vauclin, M., Vierira, S.R., Bernard, R. & Hatfield, J.L. (1982) Variability of surface temperature along two transects of a bare soil. *Water Resources Research*, **18**, 1677–1686.

Veregin, H (1994) Integration of simulation modeling and error propagation for the buffer operation in GIS. *Photogrammetric Engineering and Remote Sensing*, **60**(4), 427–435.

Vieira, S.R., Nielsen, D.R. & Biggar, J.W. (1981) Spatial variability of field-measured infiltration rate. *Soil Science Society of America Journal*, **45**, 1040–1048.

Vieux, B. (1991) Geographic informaton systems and non-point source water quality and quantity modeling. *Hydrological Processes*, **5**, 101–113.

Voss, R.F. (1988) Fractals in nature: from characterization to simulation. In: Peitgen, H.O. & Saupe, D. (eds) *The Science of Fractal Images*, 21–70. Springer-Verlag, New York, NY.

Waggoner, P.E & Stephens, G.R. (1970) Transition probabilities for a forest. *Nature*, **225**, 1160–1161.

Walsh, S.J., Lightfoot, D.R. & Butler, D.R. (1987) Recognition and assessment of error in geographic information systems. *Photogrammetric Engineering and Remote Sensing*, **53**(10), 1423–1430.

Wang, F. (1990) Improving remote sensing image analysis through fuzzy information representation. *Photogrammetric Engineering and Remote Sensing*, **56**, 1163–1169.

Ware, C. & Maggio, R.C. (1990) Integration of GIS and remote sensing for monitoring forest tree diseases in central Texas. In: *Proceedings, GIS/LIS '90, Anaheim, CA, Volume 2*, 724–732. American Society for Photogrammetry and Remote Sensing, Bethesda, MD.

Webster, R. (1973) Automatic soil-boundary location from transect data. *Mathematical Geology*, **5**, 27–37.

Webster, R. (1977) *Quantitative and Numerical Methods for Soil Survey Classification*. Oxford University Press, Oxford.

Webster, R. & Burgess, T.M. (1984) Sampling and bulking strategies for estimating soil properties in small regions. *Journal of Soil Sciences*, **35**, 127–140.

Webster, R. & Wong, I.F.T. (1969) A numerical procedure for testing soil boundaries interpreted from air photographs. *Photogrammetria*, **24**, 59–72.

Weins, J.A. & Milne, B.T. (1989) Scaling of "landscapes" in landscape ecology, or, landscape ecology from a beetle's perspective. *Landscape Ecology*, **3**, 87–96.

Welch, R. & Remillard, M.M. (1992) Integration of GPS, remote sensing, and GIS techniques for coastal resource management. *Photogrammetric Engineering and Remote Sensing*, **58**(11), 1571–1578.

Welch, R., Remillard, M.M. & Doren, R.F. (1995) GIS database development for South Florida's national parks and preserves. *Photogrammetric Engineering and Remote Sensing*, **61**(11), 1371–1381.

Welch, R., Remillard, M.M. & Slack, R.B. (1988) Remote sensing and geographic information system techniques for squatic resource evaluation. *Photogrammetric Engineering and Remote Sensing*, **54**, 177–185.

Wells, M.L. & McKinsey, D.E. (1993) The spatial and temporal distribution of lightning strikes in San Diego County, California. In: *GIS/LIS'93 Proceedings, Minneapolis, MN, Volume, 2*, 768–777. American Society for Photogrammetry and Remote Sensing, Bethesda, MD.

White, D.A. & Hofschen, M. (1996) Spatial analysis of nutrient loads in rivers and streams. In: Goodchild, M.R., Steyaert, L.T., Parks, B.O. *et al.* (eds) *GIS and Environmental Modeling: Progress and Research Issues*, 391–396. GIS World Books, Fort Collins, CO.

White, W. (1986) Modeling forest pest impacts — aided by a geographic information system in a decision support system framework. In: *Proceedings of Geographic Information Systems Workshop*, 238–248. American Society for Photogrammetry and Remote Sensing, Falls Church, VA.

Whittaker, R.H. (1956) Vegetation of the Great Smoky Mountains. *Ecological Monographs*, **26**, 1–80.

Whittaker, R.H. (1960) Vegetation of the Siskiyou Mountains, Oregon and California. *Ecological Monographs*, **30**, 279–338.

Whittaker, R.H. (1970) *Communities and Ecosystems.* Macmillan, London.

Wierenga, P.J., Hendrickx, J.M.H., Nash, M.H., Ludwig, J.A. & Daugherty, L.A. (1987) Variation of soil and vegetation with distance along a transect in the Chihuahuan Desert. *Journal of Arid Environments*, **13**, 53–63.

Wilson, J.P. (1996) GIS-based land surface/subsurface modeling: new potential for new models? *Proceedings of the Third International Conference/Workshop on Integrating GIS and Environmental Modeling.* National Center for Geographic Information and Analysis, Santa Barbara, CA. (CD-ROM.)

Wischmeier, W.H. & Smith, D.D. (1978) Predicting rainfall erosion losses. *Agricultural Handbook 537.* U.S. Department of Agriculture, Washington, D.C.

Wolter, P.T., Mladenoff, D.J., Host, G.E. & Crow, T.R. (1995) Improved forest classification in the northern lake states using multi-temporal Landsat imagery. *Photogrammetric Engineering and Remote Sensing*, **61**(9), 1129–1143.

Wynne, R.H. & Lillesand, T.M. (1993) Satellite observation of lake ice as a climate indicator: initial results from statewide monitoring in Wisconsin. *Photogrammetric Engineering and Remote Sensing*, **59**(6), 1023–1031.

Yates, S.R. & Warrick, A.W. (1987) Estimating soil water content using cokriging. *Soil Science Society of America Journal*, **51**, 23–30.

Yates, S.R. & Yates, M.V. (1990) *Geostatistics for Waste Management: A User's Manual for the GEOPACK (Version 1.0) Geostatistical Software System.* Rep. No. EPA/600/8-90/004. U.S. Environmental Protection Agency, Ada, OK.

Yates, S.R., Warrick, A.W., Matthias, A.D. & Musil, S. (1988) Spatial variability of remotely-sensed surface temperatures at field scale. *Soil Science Society of America Journal*, **52**, 40–45.

Young, T.N., Eby, J.R., Allen, H.L., Hewitt III, M.J. & Dixon, K.R. (1987) Wildlife habitat analysis using Landsat and radiotelemetry in a GIS with application to spotted owl preference for old growth. In: *Proceedings, GIS '87, San Francisco, CA, Volume 2*, 595–600. American Society for Photogrammetry and Remote Sensing, Falls Church, VA.

Zadeh, L.A. (1965) Fuzzy sets. *Information and Control*, **8**, 338–353.

Zainal, A.J.M., Dalby, R.H. & Robinson, I.S. (1993) Monitoring marine ecological changes in the east coast of Bahrain with Landsat TM. *Photogrammetric Engineering and Remote Sensing*, **59**(3), 415–421.

Zavala, I. & Garcia, R. (1992) Get the point? GPS ground control for satellite images. *GPS World*, **3**(9), 34–37.

Zhu, Z. & Evans, D.L. (1994) U.S. forest types and predicted percent forest cover from AVHRR data. *Photogrammetric Engineering and Remote Sensing*, **60**(5), 525–532.

Zsilinszky, V.G. (1966) *Photographic Interpretation of Tree Species in Ontario.* Ontario Department of Lands and Forests, Ontario.

Index

Note: Page references in *italics* refer to Figures; those in **bold** refer to Tables